Climate Change Management

Series Editor

Walter Leal Filho, International Climate Change Information and Research Programme, Hamburg University of Applied Sciences, Hamburg, Germany

The aim of this book series is to provide an authoritative source of information on climate change management, with an emphasis on projects, case studies and practical initiatives – all of which may help to address a problem with a global scope, but the impacts of which are mostly local. As the world actively seeks ways to cope with the effects of climate change and global warming, such as floods, droughts, rising sea levels and landscape changes, there is a vital need for reliable information and data to support the efforts pursued by local governments, NGOs and other organizations to address the problems associated with climate change. This series welcomes monographs and contributed volumes written for an academic and professional audience, as well as peer-reviewed conference proceedings. Relevant topics include but are not limited to water conservation, disaster prevention and management, and agriculture, as well as regional studies and documentation of trends. Thanks to its interdisciplinary focus, the series aims to concretely contribute to a better understanding of the state-of-the-art of climate change adaptation, and of the tools with which it can be implemented on the ground.

Notes on the quality assurance and peer review of this publication

Prior to publication, the quality of the works published in this series is double blind reviewed by external referees appointed by the editor. The referees are not aware of the author's name when performing the review; the referees' names are not disclosed.

Moharana Choudhury · Gopal Dixit ·
Sushobhan Majumdar
Editors

Sustainable Synergy: Harnessing Ecosystems for Climate Resilience

Springer

Editors
Moharana Choudhury
Environmental Research and Management Division
Voice of Environment (VoE)
Guwahati, Assam, India

Gopal Dixit
Department of Botany
Upadhi PG College
Mahatma Jyotiba Phule Rohilkhand University
Pilibhit, Uttar Pradesh, India

Sushobhan Majumdar
Department of Geography
Pandit Raghunath Murmu Smriti Mahavidyalaya
Bankura, West Bengal, India

ISSN 1610-2002 ISSN 1610-2010 (electronic)
Climate Change Management
ISBN 978-3-031-77956-5 ISBN 978-3-031-77957-2 (eBook)
https://doi.org/10.1007/978-3-031-77957-2

This Springer imprint is published by the registered company Springer Nature Switzerland AG
The registered company address is: Gewerbestrasse 11, 6330 Cham, Switzerland

The editors of this book dedicate the book to the Great Almighty and their family members.

Foreword

Climate change is the meta challenge of our time, which affects almost all part of our daily life and the society, in this publication described as ecosystems. To understand and respond to its challenges through the stimulation of resilience has become a mainstreamed approach in urban planning, governance, architecture and many other fields. This book provides a valuable contribution to the academic field by focussing on these ecosystems and how they can be activated for climate resilience.

The idea and outcome of this publication is extremely relevant because it acknowledges in its structure and content the systemic understanding of ecological resources and at the same time addressing climate resilience with an understanding that emphasizes also that climate is a system in itself and resilience a systemic approach. Not shying away from the inevitable complexity that these systems bring forward is an outstanding achievement of this publication. Leaving behind in many contributions traditional pathways of linear cause-effect logic and instead approaching the different ecosystems with an integrated, cross-disciplinary and holistic mindset is a scientifically proven way to deal with complexity, unpredictability, uncertainty and the constant change that is inherent to any system. This is especially true for any system that is influenced by climate-change, which is increasing in its acceleration and effects. With this increasing influence of climate change the levels of complexity in the described systems are increasing. We can feel this in by various disruption of our daily life for example through severe weather events which are now happening much more frequently.

In the spirit of UNESCO's sustainability science approach we can see in this publication how this more advanced methodology that take more effort and rigor can bring forward valuable results and new knowledge that can be beneficial to our human response to climate-change across a variety of different disciplines. The different ecosystems that are examined range from biodiversity, food, water security, communities to technologies. The book shows that all are relevant for climate-resilience in their own way and the broad selection of different ecosystems shows already that to fully understand resilience and promote it is no solution to carve out small parts of the systems, but rather a wide understanding of different topics and especially making an effort to understand how they are connected, interrelated and influencing each other is a smarter way to approach this complex topic.

The multidisciplinary content of this publication together with specific examples and theoretical approaches provide an important contribution to the discourse.

Dr. Matthias Ripp
Senior Heritage Manager
Brandenburg University of
Technology Cottbus
Cottbus, Germany
matthiasripp@posteo.de

Preface

In an era where the impacts of climate change are increasingly evident, the need for innovative and holistic approaches to building resilience has never been more pressing. *Sustainable Synergy: Harnessing Ecosystems for Climate Resilience* emerges as a timely contribution to the ongoing dialogue on how humanity can adapt to and mitigate the multifaceted challenges posed by a warming planet. This book is rooted in the belief that ecosystems, when managed sustainably and thoughtfully, hold the key to enhancing our resilience against climate change. From the forests that regulate our climate to the wetlands that shield us from storms, nature provides a wealth of solutions that can be harnessed to protect and sustain communities around the globe. However, the potential of these natural systems is often underutilized or overlooked in the planning and implementation of climate resilience strategies. The chapters in this volume offer a comprehensive exploration of ecosystem-based approaches to climate resilience. Contributors from diverse fields—ranging from ecology and environmental science to urban planning and public policy—provide insights into how ecosystems can be leveraged to build resilient societies. By integrating scientific knowledge with practical case studies, this book not only highlights the importance of ecosystems but also offers actionable frameworks and strategies for their effective use. "Sustainable Synergy" is not just a collection of academic discussions; it is a call to action. It invites policymakers, researchers, practitioners, and community leaders to rethink traditional approaches to climate resilience and to recognize the critical role that ecosystems play in our survival. In doing so, it seeks to inspire a new paradigm in which human development and environmental stewardship go hand in hand, leading to a more sustainable and resilient future. We hope that this book will serve as a valuable resource and a source of inspiration for all those committed to addressing the challenges of climate change through the power of nature.

Guwahati, India Moharana Choudhury
Pilibhit, India Gopal Dixit
Bankura, India Sushobhan Majumdar

Acknowledgements

Writing a book is more complex than we thought and more rewarding than we imagined. The editors would like to thank Springer Nature for their constant support for this edited book volume. Support from acquisition editor Quiao Shu and production editor Mr. Ashok Arumairaj is highly appreciated during the entire publication process. This endeavour would not have been possible without the active participation of our esteemed contributing authors and their quality chapters, and it is a matter of pride for all the editors of this book. Through their contributed chapters, their support for the broader scientific dissemination of the knowledge and aspects on ecosystem services and climate resilience brings new audiences to know the challenges in resilient climatic conditions and sustainable synergy.

The editors are thankful for vital support has been obtained from the NGOs, Environment Conservation Organization (ECO) from UP state and "Voice of Environment" (A Scientific and Environmental Research Organization) for their support and encouragement during the book editing and compilation process.

We, the editors would like to extend our gratitude to all the esteemed contributing authors across the world, our colleagues, research supervisors, collaborators, and scholars for their direct or indirect help.

We would like to acknowledge the vital contributions of all the researchers worldwide involved in this project.

At last but not least the editors of this book dedicate the book to the Great Almighty and their family members.

About This Book

Sustainable Synergy: Harnessing Ecosystems for Climate Resilience offers a groundbreaking exploration of how integrating natural ecosystems into climate strategies can enhance environmental sustainability and bolster resilience. This insightful book delves into the concept of ecosystem-based adaptation, showcasing real-world examples where the protection and restoration of natural habitats have led to effective climate solutions.

The author, a leading expert in environmental science, examines a range of ecosystems—from forests and wetlands to coastal areas—and illustrates their crucial roles in mitigating climate change impacts. Through a combination of case studies, scientific research, and practical guidance, the book provides readers with actionable strategies for leveraging nature's own mechanisms to address climate challenges.

Ideal for policymakers, environmentalists, and anyone passionate about climate action, "Sustainable Synergy" not only highlights the urgent need for ecosystem-based approaches but also offers a hopeful vision of a collaborative future where humanity and nature thrive together.

Contents

Editors and Contributors

About the Editors

Moharana Choudhury is an Indian researcher and environmentalist. He is currently a Research Coordinator in the Environmental Research and Management Division of "Voice of Environment" (A Scientific and Environmental Research Organization), Guwahati, Assam, India. Choudhury has worked on various scientific projects in various Government departments and has more than 13 years of research and working experience in environmental research and services. He began his career as a Research Intern at the National Institute of Disaster Management in New Delhi, followed by the North Eastern Space Application Center (Department of Space, Government of India/ISRO Unit) as a Junior Research Fellow. Choudhury has authored or edited ten (10) books with internationally reputed publication houses and more than thirty-six (36) articles in national and internationally reputed journals, proceedings, and more than (30) thirty book chapters. He has written many articles and scientific reports in popular magazines. He is working as an editor for various prestigious international publishing houses worldwide, such as Elsevier, Royal Society of Chemistry, Springer, Wiley, CRC Press/ Routledge group, Apple Academic Press, Cambridge Scholars Publishing etc. Choudhury is working on various environmental conservation activities in Assam and the country. He has worked on several initiatives to integrate environmentally sustainable projects for iconic heritage places, wetlands, and ecotourism spots in Assam, India, especially for cleanliness and beating single-use plastic pollution missions. He contributes to various environmental research, management, social media mass awareness, and research studies. He is also well known in India for his efforts to implement environmental education and safeguard the privileges of environmental degree holders in India. He has completed his education as an MSc, PGDGIS, PGDEM, and PGCGARD from Gurukul Kangri Deemed to be University, Haridwar, Uttrakhand, North Eastern Hill University (NEHU), Shillong Meghalaya, Tezpur University, Assam, and National Institute of Rural Development (NIRDPR) Hyderabad, India, respectively.

Dr. Gopal Dixit is the Head of the Department of Botany at Upadhi PG College, Pilibhit, India, under MJP Rohilkhand University. With a Ph.D. in Botany, Dr. Dixit has a distinguished career in teaching and research, focussing on Botany, Ethnobotany, Plant Pathology, Plant Physiology, Nematology and Environmental Sciences. He is currently pursuing a D.Sc. in Botany. Dr. Dixit has authored several text books, and numerous research papers, with notable contributions to the fields of plant sciences and environmental conservation. His work extends to various national and international platforms, where he has presented research on topics such as the impact of Fertilizer factory effluents on the plant growth and the ethnobotanical practices of tribal communities in the Indo–Nepal sub Himalayan terai region. In addition to his academic and research endeavours, Dr. Dixit h as secured patents in the field of agricultural technology, including innovations in precision farming and weed control devices. His commitment to the sustainable management of natural resources is evident in his publications and projects, which often explore the intersection of traditional knowledge and modern scientific practices. Dr. Dixit also actively engages in public education through his YouTube channel, "Study with GD Sir", where he shares his expertise with a broader audience.

Dr. Sushobhan Majumdar is an Assistant Professor in the Department of Geography at Pandit Raghunath Murmu Smriti Mahavidyalaya (Affiliated to Bankura University, India). He is an applied geographer who holds a doctorate in geography from the Department of Geography at the Faculty of Science at Jadavpur University in Kolkata and a postgraduate degree from Presidency University in Kolkata. Previously he was employed by the Government of West Bengal as a 'Group A Officer' in the "Minority Affairs and Minority Education Department". He is presently engaged in research on the design and development of environmental geographies. Urban geography, social geography, regional development and planning, remote sensing and GIS, and geography of transport are among his main areas of study. Additionally, he is working on different development projects for the Government of West Bengal's Socio Economic Planning Unit of the Kolkata Metropolitan Development Authority (KMDA). He has more than 50 research papers published in a variety of reputable journals, proceedings, and edited volumes both nationally and internationally. He has already authored 7 edited volumes for Springer Nature, Elsevier and Routledge Group on urban-related issues. He is now editor of various renowned publishing houses in the world like Elsevier, Royal Society of Chemistry, Springer, CRC Press, AAP Press, Wiley, Cambridge Scholars Publishing etc. He has presented papers at more than 30 national and international seminars and conferences, both in India and abroad. He gave guest or special talks as part of numerous national and regional programs. He holds memberships in the Geographical Society of India, the Regional Science Association of India (RSAI), the Institute of Indian Geographers (IIG), and the National Association of Geographers of India (NAGI).

Contributors

Khalid Abbas Department of Zoology, Wildlife and Fisheries, University Of Agriculture, Faisalabad, Pakistan

Godfred Addai Department of Planning, Planning Institute of Australia, Barton, Australia

Moses Kwame Aidoo Cocoa Research Institute of Ghana, New Tafo-Akim, Eastern Region, Ghana

Fatima Amin Department of Zoology, Wildlife and Fisheries, University Of Agriculture, Faisalabad, Pakistan

Shivam Attri Department of Environmental Sciences, Graphic Era Hill University, Dehradun, Uttarakhand, India

Ayantika Banerjee Federal Institute for Material Research and Testing, BAM, Federal Ministry for Economic Affairs and Climate Protection, Berlin, Germany

Mukul Machhindra Barwant Department of Botany, Sanjivani Arts Commerce and Science College Kopargaon, Kopargaon, Maharashtra, India

Muqadas Batool Department of Botany, (Govt Graduate College Layyah) Bahauddin Zakariya University, Multan, Pakistan

Shumaila Batool Department of Zoology, Wildlife and Fisheries, University of Agriculture, Faisalabad, Pakistan

Donizete Beck Insper Institute of Education and Research, Sao Paulo, Brazil

N. Bharathy VIT School of Agricultural Innovations and Advanced Learning, Vellore Institute of Technology (VIT), Vellore, India;
School of Bio Sciences and Technology, Vellore Institute of Technology (VIT), Vellore, India

Laxmi Kant Bhardwaj Amity Institute of Environmental Toxicology, Safety & Management, Amity University, Noida, Uttar Pradesh, India

Abhilash Bhattacharjee Department of Botany, D.H.S.K. College, Dibrugarh, Assam, India

Mihika Bhattacharjee Department of Biotechnology, Bengal Institute of Technology, Kolkata, West Bengal, India

Parnali Bhattacharya Department of Genetics and Plant-Breeding (Agriculture), University of Calcutta, Kolkata, West Bengal, India

Soutrik Bhattacharya Department of Biotechnology, Bengal Institute of Technology, Kolkata, West Bengal, India

Ambarish Bhuyan Department of Botany, D.H.S.K. College, Dibrugarh, Assam, India

Saptaswa Chakraborty Department of Biotechnology, Bengal Institute of Technology, Kolkata, West Bengal, India

Avnish Chauhan Department of Environmental Sciences, Graphic Era Hill University, Dehradun, Uttarakhand, India

Moharana Choudhury Environmental Research and Management Division, Voice of Environment (VoE), Guwahati, Assam, India

Banani Ray Chowdhury Department of Biotechnology, Bengal Institute of Technology, Kolkata, West Bengal, India;
Baranagar Baghajatin Social Welfare Organisation, Kolkata, West Bengal, India;
World Education & Culture Trust, Kolkata, West Bengal, India

Ratan Chowdhury Department of Botany, Rangapara College, Rangapara, Assam, India

Einstein Mariya David VIT School of Agricultural Innovations and Advanced Learning, Vellore Institute of Technology (VIT), Vellore, India;
School of Bio Sciences and Technology, Vellore Institute of Technology (VIT), Vellore, India

Sukhendu Dey The University of Burdwan, Burdwan, West Bengal, India

Gopal Dixit Department of Botany, Upadhi PG College (MJP Rohilkhand University), Pilibhit, UP, India

Minahal Fatima Department of Zoology, Wildlife and Fisheries, University Of Agriculture, Faisalabad, Pakistan

Saiyyeda Firdous VIT School of Agricultural Innovations and Advanced Learning, Vellore Institute of Technology (VIT), Vellore, India

Mariana Flores-García Directora del Instituto Nacional de Arquitectura y Urbanismo, Universidad De La Salle Bajío, Leon, Mexico

Zacharias Frontistis Department of Chemical Engineering, University of Western Macedonia, Kozani, Greece

Rekha Goswami Department of Environmental Sciences, Graphic Era Hill University, Dehradun, Uttarakhand, India

Lawrence Guodaar Department of Geography and Rural Development, Kwame Nkrumah University of Science and Technology, Kumasi, Ghana

Md. Yeamin Hossain Department of Fisheries, Faculty of Fisheries, University of Rajshahi, Rajshahi, Bangladesh

Harshita Jain Amity Institute of Environmental Sciences, Amity University, Noida, Uttar Pradesh, India

Pankaj Kanti Jodder Department of Geography and Earth Science, University of North Carolina at Charlotte, Charlotte, USA

Kamal Kant Joshi Department of Environmental Sciences, Graphic Era Hill University, Dehradun, Uttarakhand, India

Anmoldeep Kaur Department of Agriculture, Royal Agricultural University, Cirencester, Gloucestershire, UK

Parvez Alam Khan Department of Management and Humanities, Universiti Teknologi PETRONAS, Seri Iskandar, Malaysia

P. Kowsalya VIT School of Agricultural Innovations and Advanced Learning, Vellore Institute of Technology (VIT), Vellore, India

T. R. Kumaraswamy Centre for Sustainability, School of Law, Governance and Public Policy, Chanakya University, Bengaluru, India

Pankaj Kumar Department of Veterinary Physiology, College of Veterinary and Animal Science, RAJUVAS, Bikaner, Bikaner, India

Prasann Kumar Department of Agronomy, School of Agriculture, Lovely Professional University, Phagwara, Punjab, India

Kehu Li College of Life Sciences, Guizhou University, Guiyang, Guizhou, China

Sushobhan Majumdar Department of Geography, Pandit Raghunath Murmu Smriti Mahavidyalaya, Bankura, West Bengal, India

Rwitabrata Mallick Amity School of Life Science, Amity University Madhya Pradesh, Gwalior, India

Mohammad Reza Morshedloo Department of Horticultural Sciences, Faculty of Agriculture, University of Maragheh, Maragheh, Iran

Murad Muhammad Xinjiang Key Laboratory of Biodiversity Conservation and Application in Arid Lands, Xinjiang Institute of Ecology and Geography, Chinese Academy of Sciences, Urumqi, People's Republic of China;
State Key Laboratory of Desert and Oasis Ecology, Key Laboratory of Ecological Safety and Sustainable Development in Arid Lands, Xinjiang Institute of Ecology and Geography, Chinese Academy of Sciences, Urumqi, People's Republic of China;
University of Chinese Academy of Sciences, Beijing, People's Republic of China;
State Key Laboratory of Desert and Oasis Ecology, Key Laboratory of Ecological Safety and Sustainable Development in Arid Lands, Xinjiang Institute of Ecology and Geography, Chinese Academy of Sciences, Beijing, China;
University of Chinese Academy of Sciences, Beijing, China;
Xinjiang Key Laboratory of Biodiversity Conservation and Application in Arid Lands, Xinjiang Institute of Ecology and Geography, Chinese Academy of Sciences, Beijing, China

Anindita Pal Department of Geography, The Maharaja Sayajirao University of Baroda, Gujarat, India

Debopriyo Pal Department of Biotechnology, Bengal Institute of Technology, Kolkata, West Bengal, India

T. Parthasarathi VIT School of Agricultural Innovations and Advanced Learning, Vellore Institute of Technology (VIT), Vellore, India

Sandeep Poddar Lincoln University College, Petaling Jaya, Selangor, Malaysia

Shivani Ponappa School of Biotechnology, University of Leads, Leads, UK

Rupam Priya Department of Bioscience and Biotechnology, Banasthali Vidyapith, Vanasthali, Rajasthan, India

Sanju Purohit Environmental/Ecological Studies and Sustainability, Akamai University, Kamuela, HI, USA

Ravi Rajak Department of Economics, The University of Bristol, Bristol, UK

Prangya Rath Amity Institute of Environmental Sciences, Amity University, Noida, Uttar Pradesh, India

Nimai Chandra Saha Post Graduate Department of Zoology, Bidhannagar College, Kolkata, West Bengal, India

Shubhajit Saha Fishery and Ecotoxicology Research Laboratory, Department of Zoology, The University of Burdwan, Burdwan, India

Aanchal Saklani Department of Environmental Sciences, Graphic Era Hill University, Dehradun, Uttarakhand, India

Induja Samanta Department of Biotechnology, Bengal Institute of Technology, Kolkata, West Bengal, India

Bipin Sati Department of Science and Technology, Govt. of Uttarakhand, Dehradun, Uttarakhand, India

Kamalesh Sen Department of Environmental Science, Syamsundar College, Burdwan, West Bengal, India

Shreya Sen Department of Biotechnology, Bengal Institute of Technology, Kolkata, West Bengal, India

Bhawana Sharma Department of Bioscience and Biotechnology, Banasthali Vidyapith, Vanasthali, Rajasthan, India

Prernaa Sharma Department of Environmental Sciences, Graphic Era Hill University, Dehradun, Uttarakhand, India

Joginder Singh Department of Botany, Nagaland University, Nagaland, India

Daniel D. Snow School of Natural Resources and Nebraska Water Center, part of the Daugherty Water for Food Global Institute, University of Nebraska, Lincoln, NE, USA

Shalini Thakur Circular Materials, Toronto, ON, Canada

Atul K. Tiwari Department of Geography, The Maharaja Sayajirao University of Baroda, Gujarat, India

Sana Ullah Department of Agricultural Leadership, Education & Communication College of Agricultural & Environmental Sciences, University of Georgia, Athens, GA, USA

Shahid Ullah Department of Botany, University of Peshawar, Peshawar, Pakistan

Eric D. van Hullebusch Université Paris Cité, Institut de Physique du Globe de Paris, CNRS, Paris, France

Abdul Wahab Shanghai Center for Plant Stress Biology, CAS Center for Excellence in Molecular Plant Sciences, Chinese Academy of Sciences, Shanghai, People's Republic of China

Ajar Nath Yadav Department of Genetics, Plant Breeding and Biotechnology, Dr. Khem Singh Gill Akal College of Agriculture, Eternal University, Baru Sahoib, Sirmour, Himachal Pradesh, India

Poonam Yadav ICAR-Indian Institute of Maize Research, Ludhiana, Punjab, India

Purvi Zaveri Biocare Research (I) Pvt. Ltd., Sustainability and Innovation Co Chair, YPFP, Tokyo, Japan

Anran Zheng AECOM Technical Services Inc., Dallas, TX, USA

Abbreviations

ACIA	Arctic Climate Impact Assessment
AEZ	Agroecological Zone
AI	Artificial Intelligence
ANN	Artificial Neural Network
ARCPATH	Arctic Climate Pathways
ASR	Aquifer Storage and Recharge
BECCS	Bioenergy with carbon capture and storage
BERI	Bioclimatic Ecosystem Resilience Index
BMP	Best Management Practice
CAP	Common Agricultural Policy
CbA	Community-Based Adaptation
CBD	Convention on Biological Diversity
CBNRM	Community-Based Natural Resource Management
CCLL	Coastal City Living Lab
CCSE	Climate Change and Sustainable Energy
CES	Cultural Ecosystem Services
CICES	Common International Classification of Ecosystem Services
CNN	Convolutional Neural Network
CO_2	Carbon dioxide
COP	Conference of Parties
CSA	Climate-Smart Agriculture
CoSA	Community-Supported Agriculture
CUSPRA	Cusp Resilience Assessment
DNA	Deoxyribonucleic acid
EbA	Ecological Balance Assessment
EbA	Ecosystem-based adaptation
EBA	Ecosystem-Based Approach
ES	Ecosystem Service
EUCCLN	Enhanced Urban Carrying Capacity Load Number
EU	European Union
GIS	Geographic Information System

GMO	Genetically Modified Organism
ICCAT	International Commission for the Conservation of Atlantic Tunas
ICT	Information and Communications Technology
ICZM	Integrated Coastal Zone Management
IFAD	International Fund for Agriculture Development
ILK	Indigenous and Local Knowledge
ILM	Integrated landscape management
ILP	Indigenous Local People
IOTC	Indian Ocean Tuna Commission
IoT	Internet of Things
IPCC	Intergovernmental Panel on Climate Change
IP	Indigenous People
IPM	Integrated Pest Management
ISO	International Organization for Standardization
IUCN	International Union for Conservation of Nature
IWRM	Integrated Water Resources Management
LiDAR	Light Detection and Ranging
MEA	Million Ecosystem Assessment
MPA	Marine Protected Area
NbS	Nature Based Solutions
NbT	Nature-based technologies
NCONF	National Centre for Organic and Natural Farming
NGO	Non-Governmental Organization
NPOP	National Programme for Organic Production
PA	Protected Area
PES	Payments for Ecosystem Services
PRA	Participatory Rural Appraisal
R&D	Research and Development
RNN	Recurrent Neural Network
SbT	Smart-Based Technologies
SCORE	Smart Control of Climate Resilience
SDGs	Sustainable Development Goals
SEA	Strategic Environmental Assessment
SIA	Systems Innovation Approach
SLM	Sustainable Land Management
SLM	Sustainable Livestock Management
SMP	Standardized Monitoring Protocol
SSA	Sub Saharan Africa
SSP	Shared Socioeconomic Pathway
STP	Sewage Treatment Plant
SWM	Sustainable Water Management
TEK	Traditional Ecological Knowledge
UAV	Uncrewed Aerial Vehicle
UN	United Nations
UNCCD	United Nations Convention to Combat Desertification

UNEP	United Nations Environment Programme
UNFCCC	United Nations Framework Convention on Climate Change
VRIP	Vulnerability-Resilience Indicator Prototype
WCPFC	Western and Central Pacific Fisheries Commission

Part I
Ecosystem Services Unveiled: Building Climate Resilience

Chapter 1
Introduction to Ecosystem Services for Climate Resilience: A Theoretical Framework

Gopal Dixit, Sushobhan Majumdar, and Moharana Choudhury

Abstract Climate change poses unprecedented challenges to ecosystems and human societies worldwide, necessitating innovative approaches to enhance resilience. Ecosystem services, the benefits humans derive from ecosystems, are crucial in bolstering resilience against climate impacts. This paper introduces a theoretical framework for understanding how ecosystem services contribute to climate resilience. It synthesizes existing literature on ecosystem services and resilience, highlighting key concepts such as provisioning, regulating, cultural, and supporting services. The framework integrates ecological, social, and economic dimensions, emphasizing the interconnectedness between natural and human systems. This framework aims to inform policy and management strategies for climate adaptation by elucidating the mechanisms through which ecosystem services enhance resilience. Case studies illustrating the framework's application underscore its practical utility in diverse socio-ecological contexts. Ultimately, this theoretical foundation provides a basis for further research and action to promote sustainable development in the face of climate change.

Keywords Ecosystem services · Climate resilience · Climate adaptation · Sustainable development

G. Dixit
Department of Botany, Upadhi PG College (MJP Rohilkhand University), Pilibhit, UP 262001, India

S. Majumdar
Department of Geography, Pandit Raghunath Murmu Smriti Mahavidyalaya, Bankura, West Bengal, India

M. Choudhury (✉)
Environmental Research and Management Division, Voice of Environment (VoE), Guwahati, Assam 781034, India
e-mail: moharanachoudhury@gmail.com

M. Choudhury et al. (eds.), *Sustainable Synergy: Harnessing Ecosystems for Climate Resilience*, Climate Change Management,
https://doi.org/10.1007/978-3-031-77957-2_1

Introduction

The products or advantages given by the ecosystems are commonly known as ecosystem services, ESs. Biodiversity encompasses the different life forms on our planet that play a crucial role in maintaining the ecosystem's stability and the climate's resilience. Additionally, it provides essential ecosystem services that support the prosperity of all life forms (Choudhury et al. 2021). Human exercises, such as habitat damage, pollution, climate change, and invasion of obtrusive species, have promoted unprecedented paces of biodiversity loss (Gogoi et al. 2024a, b). Conservation efforts and restoration initiatives have gained momentum due to the earnest need to address this emergency. The products or advantages given by the ecosystems are commonly known as ecosystem services, ESs. Biodiversity encompasses the different life forms on our planet that play a crucial role in maintaining the ecosystem's stability and the climate's resilience.

Additionally, it provides essential ecosystem services that support the prosperity of all life forms (Choudhury et al. 2021). Human exercises, such as habitat damage, pollution, climate change, and invasion of obtrusive species, have promoted unprecedented paces of biodiversity loss (Gogoi et al. 2024a, b). Conservation efforts and restoration initiatives have gained momentum worldwide because of the earnest need to address this emergency.

Ecosystem services restoration includes conscious actions to invert or mitigate the effects of human exercises on natural climate. Restoration of ecosystems involves deliberate actions to reverse or reduce the impact of human actions on natural climatic conditions. These efforts encompass a variety of activities, including habitat restoration, rehabilitation, reforestation, species reintroduction, and the creation of protected areas. Its chief motive is to enhance climate resilience through ecosystem services (Kumar et al. 2020). It also promotes species recovery, leading to genetic diversity.

On the other hand, ecosystem conservation focuses on protecting and managing natural biodiversity hotspots and critical habitats. Measures of ecosystem conservation include the foundation of safeguarded regions, species conservation programs, and the enforcement of regulations to prevent habitat destruction and the exploitation of living beings. Additionally, conservation efforts commonly involve community engagement and partnerships with native people groups and nearby networks recognizing their customary information and stewardship practices (Kumar and Choudhury 2021).

Even with developing familiarity with the importance of biological systems in climate resilience, significant challenges still need to be addressed. These include inadequate funding and ecological resources, policy gaps, conflicting interests among policymakers and stakeholders, and limited monitoring and law enforcement capacity. Moreover, the interconnected nature of ecosystem services requires coordinated action at nearby local, regional, public, and worldwide levels, posing governance and coordination challenges for climate resilience.

Advancing in ecological technologies, such as remote sensing, GIS mapping, and DNA sequencing, may offer new tools for monitoring ecosystems and accessing climate resilience. Moreover, growing public awareness and engagement, coupled with increasing recognition of the economic values of natural resources, are driving momentum for climate resilience. Multilateral initiatives, such as the Convention on Biological Diversity (CBD) and the Sustainable Development Goals (SDGs), give structure for international collaboration and goal setting in climate resilience.

Ecosystem services underscore the critical role of biodiversity in maintaining ecological balance and keeping up with the biological equilibrium of biological systems. The restoration of biodiversity and the protection of regular assets are vital components of global efforts to achieve sustainability and mitigate degradation impacts in climatic conditions (Gorman et al. 2023). By safeguarding biodiversity, we can ensure the continued provision of ecosystem services, support livelihoods, and enhance climate resilience to environmental changes.

Ecosystem services directly or indirectly contribute to the quality of human life. Practically, it provides food and water. On the other hand, it plays a role in climate resilience. As indicated by the Millennium Ecosystem Assessment (MA 2005), four classifications of biological system administrations have been identified, which are provisioning (food and water), regulating (carbon sequestration and water decomposition), supporting (soil formation and seed dispersal), and cultural (recreation and inspirational). Water purification, soil formation, crop pollination, and pest control are ecosystem services. Ecosystems provide food, water, timber, and air purification; soil creation and pollination are the best examples of ecosystem services. Several human actions impact healthy ecosystems by eroding biodiversity (Smith et al. 2021).

Ecosystem services restoration includes conscious actions to invert or mitigate the effects of human exercises on natural climate. Restoration of ecosystems involves deliberate actions to reverse or reduce the impact of human actions on natural climatic conditions. These efforts encompass a variety of activities, including habitat restoration, rehabilitation, reforestation, species reintroduction, and the creation of protected areas. Its chief motive is to enhance climate resilience through ecosystem services (Kumar et al. 2020). It also promotes species recovery, leading to genetic diversity.

On the other hand, ecosystem conservation focuses on protecting and managing natural biodiversity hotspots and critical habitats. Measures of ecosystem conservation include the foundation of safeguarded regions, species conservation programs, and the enforcement of regulations to prevent habitat destruction and the exploitation of living beings. Additionally, conservation efforts commonly involve community engagement and partnerships with native people groups and nearby networks recognizing their customary information and stewardship practices (Kumar and Choudhury 2021). Even with developing familiarity with the importance of biological systems in climate resilience, significant challenges still need to be addressed. These include inadequate funding and ecological resources, policy gaps, conflicting interests among policymakers and stakeholders, and limited monitoring and law enforcement capacity. Moreover, the interconnected nature of ecosystem services requires coordinated action at nearby local, regional, public, and worldwide levels, posing governance and coordination challenges for climate resilience.

Advancing in ecological technologies, such as remote sensing, GIS mapping, and DNA sequencing, may offer new tools for monitoring ecosystems and accessing climate resilience. Moreover, growing public awareness and engagement, coupled with increasing recognition of the economic values of natural resources, are driving momentum for climate resilience. Multilateral initiatives, such as the Convention on Biological Diversity (CBD) and the Sustainable Development Goals (SDGs), give structure for international collaboration and goal setting in climate resilience.

Ecosystem services underscore the critical role of biodiversity in maintaining ecological balance and keeping up with the biological equilibrium of biological systems. The restoration of biodiversity and the protection of regular assets are vital components of global efforts to achieve sustainability and mitigate degradation impacts in climatic conditions (Gorman et al. 2023). By safeguarding biodiversity, we can ensure the continued provision of ecosystem services, support livelihoods, and enhance climate resilience to environmental changes.

Ecosystem services directly or indirectly contribute to the quality of human life. Practically, it provides food and water. On the other hand, it plays a role in climate resilience. As indicated by the Millennium Ecosystem Assessment (MA 2005), four classifications of biological system administrations have been identified, which are provisioning (food and water), regulating (carbon sequestration and water decomposition), supporting (soil formation and seed dispersal), and cultural (recreation and inspirational). Water purification, soil formation, crop pollination, and pest control are ecosystem services. Ecosystems provide food, water, timber, and air purification; soil creation and pollination are the best examples of ecosystem services. Several human actions impact healthy ecosystems by eroding biodiversity (Smith et al. 2021).

Literature Review

Ecosystem services are related to the natural situations or activities that promote the welfare of individuals or social well-being. Individuals can significantly benefit directly or indirectly from ecosystem services, e.g., pollinating different crops by honey bees and other flies to produce specific edible products (Birkhofer et al. 2015). Environmental change is influencing ecosystem productivity in unprecedented ways.

Ecosystem services connect Nature and society for the well-being of humans and economic development (Xue et al. 2012). Massive impact of changes in land cover and land use have been noticed (Wang et al. 2021; Kumar et al. 2018; Shukla et al. 2018, 2020; Peters et al. 2019; Wu et al. 2020). Conversion in land use and land cover can deteriorate the ecosystem services globally (Nelson et al. 2010). The validity of the biological system cannot be founded only on human interventions (Xia et al. 2021). As far as ecosystem services are concerned, urban ecosystems is an urgent area of ecosystem services research, as almost 50% of the population resides in metropolitan areas. A study of MEA reveals that about 60% of the global ecosystem services have been threatened or not used correctly (MA 2005). Environment services are considered one of the vital aspects of land use arranging and ecological environment (Daily et al. 2009; Xia et al. 2021; Costanza et al. 2017;

Sattler et al. 2018; Dade et al. 2019; Cord et al. 2017; Bennett et al. 2009; Lee and Lautenbach 2016; Li et al. 2021).

Ecosystem services contribute to climate resilience in two types. The first one is compromised, where the expansion of one ES will diminish the actuality of other ES. The second one is collaborations, where the expansion of one ES will prompt the expansion in the actuality of another ES (Han et al. 2020; Dade et al. 2019). As per (Bennett et al. 2009), compromises and cooperative energies are brought about by connections between multiple environments. In this way, environmental administrations are not viewed as autonomous, and their study is always complex (Xia et al. 2021).

Monitoring of the ecosystems could be done at local, regional, or global levels, depending upon the need for research. Regarding climate resilience, researchers suggest local-level monitoring to understand the situation better. It helps to consider suitable regional mitigation strategies, i.e., at the next step. These studies will achieve global sustainable targets (Bennett et al. 2009). A few different structures arose in the past to examine how ecosystem services contribute to climate resilience (Zhang et al. 2018). Common International Classification of Ecosystem Services (CICES) incorporates different parameters for numerous ecosystems for the natural capital. Intergovernmental Science Policy Platforms on Biodiversity and Ecosystem Services (IPBES) developed a structure for evaluating the contribution of Nature to humans. In this way, ecosystems can be gathered in 4 classes, i.e., (i) provisioning ESs, (ii) regulating ESs, (iii) cultural ESs, and (iv) supporting ESs. Provisioning ESs and regulating ESs are considered necessary as far as climate resilience factors are concerned. Provisioning ESs are water, food, wood, biofuels, etc. Precipitation, vanishing, and climatic fluctuations are the significant climatic factors for the area's production (Zhang et al. 2018).

In light of current research, regulating ESs can be explained as the total gains of biological services such as climate changes, soil fertility, floods, etc. Regulating ecosystems is considered a critical assessment for climate resilience (Bennett et al. 2009). Carbon sequestration is essential for assessing climate resilience (Carter Berry et al. 2020). It is well established that in Nature, carbon is put away in 4 unique layers: (i) in above-ground Biomass, (ii) underground Biomass,(iii) dead matter carbon, and (iv) soil organic matter (Gogoi et al. 2024a, b). As per an estimation, nearly 70% of the carbon is stored in the land ecosystem, and due to increasing CO_2 levels, its natural balance could be effective badly shortly (Delibas et al. 2021; Kumar et al. 2021; Rai et al. 2021). Different land use/land change situations are considered hotspots of future research, which helps vital policymakers form mitigation strategies and frameworks for better reach.

Socio-cultural practices with human psychological development are included in Cultural Ecosystem Services (CES), where benefits are assumed and attained due to interaction (Sen and Gucchait 2021). Supporting ESs are the fundamental and critical actions of the biological system and necessary for life on the earth, e.g., Photosynthesis, nutrient cycle, and evolutionary practices. For future research opportunities and decision-making practices, it is essential to see every biological system and its impact on climate resilience for sustainable development. This research can provide

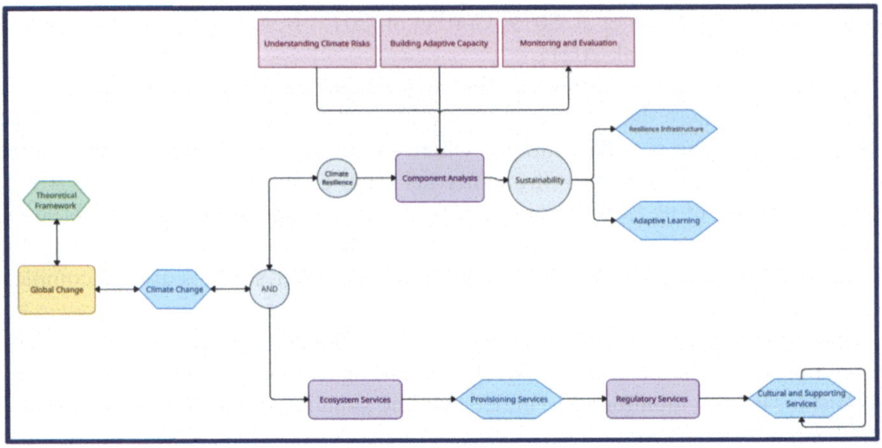

Fig. 1.1 Methodological framework

future decision-making strategies on national and global levels. This study mainly focuses on the worldwide status of ecosystem services and identifies the possible research gaps (Fig. 1.1).

Case Studies: Role of ESs in Promoting Climate Resilience

Ecosystem services always play a promotional role in strategically managing climate variations and adaptation. Ecosystems directly contribute to mitigation as they remove Carbon from the atmosphere. They played a vital role in people's adaptation to climate crises and future climate changes. Ecosystem services and climate resilience are tightly linked together. This chapter explores the possibilities of ecosystem services in mitigation and adaptation and threat to climate change at local, meso-level, regional, and global adaptations. In this way, the main focus is on provisioning ESs (food and timber) and regulating ESs (water and pest control). Regarding cultural ESs (recreation, elegance, and religious practices) and supporting ESs (primary commodity and biogeochemical cycle), they are of little adaptation importance but have ecological resilience benefits (Fig. 1.2).

Ecosystem services can result in well-managed ecosystems for present and future climate changes (Doswald et al. 2014). In developing countries, provisioning services are crucial for the adapting strategies of rural populations. Many of such communities depend upon forest products in drought conditions. Similarly, in flood situations in Peru and droughts in Tanzania, forest and wild products were extensively used to meet their needs. After the Hurricane in Honduras, local people depended on forest products to recover from adverse conditions (McSweeney 2005). Wood and wood forest products (uncultivated fruits, vegetables, Mushrooms, and forage) also

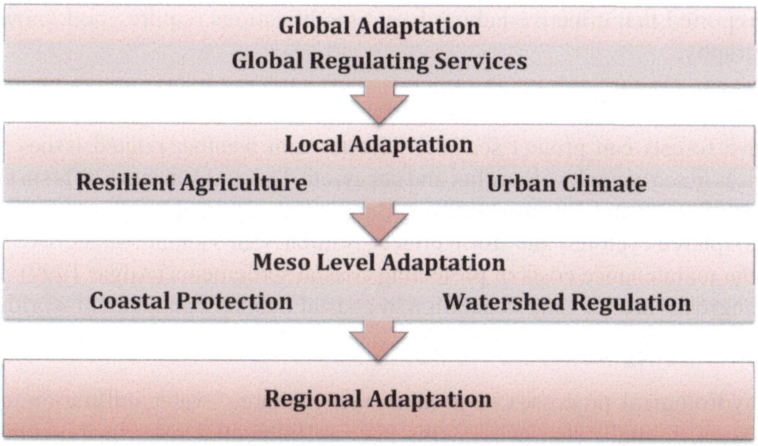

Fig. 1.2 Ecosystem services for climate adaptation

contributed generously to facing these adversities in Cameroon (Bele et al. 2011) and Bolivia (Robledo et al. 2004). In Mali, despite climate variations, compound agriculture systems associated with several crops, forage, and vegetation ensure a constant food supply (Djoudi et al. 2013). Almost similar types of results have been observed by different workers in Malawi (Fisher et al. 2010), in the Philippines (Walton et al. 2006), and in Indonesia (Liswanti et al. 2011). This study has identified many case studies for the role of different ecosystem services in promoting climate resilience in different geographical locations. These location variations result in habitat types, climate, and environment changes. In this way, ecosystem services directly or indirectly affect climate resilience globally. Some of the successful case studies are as follows:

Regulating Services in Local Agriculture

As we know, trees in and around fields can regulate cropping systems under climatic variations. Deep roots of trees can bring sufficient water and nutrients during drought conditions. On the other hand, they will improve soil fertility and check the process of soil erosion in Zambia and Malawi (Garrity et al. 2010). In Mexico, the shade of trees can maintain the temperature and humidity for better protection from storms and speedy winds in Coffee and Cacao production (Lin 2010). High tree cover, on the one hand, increases the protection of soil and, on the other hand, reduces the availability of light to crops understory (Verchot et al. 2007).

Regulating Services in City Region

Vegetation plays a crucial role in regulating local temperatures and water through evaporation, shade rain, and infiltration in cities and urban areas. As noticed in Manchester, the UK's local temperature rise can be moderated by the increase in the area's green cover (Gill et al. 2007). In Durban, South Africa, Roberts et al.

(2012) reported that effective habitat-based modifications require good knowledge of topography.

Coastal Habitat Management

Mangrove forests can protect seaside habitats from weather-related issues, hurricanes, rises in sea water levels, influx and decay, etc. Das and Vincent (2009) in Orissa, India, and Iftekhar and Takama (2008) in Bangladesh reported that mangroves effectively conquered cyclones and storm effects. Similarly, in Vietnam, mangrove forests reduce the maintenance costs of protecting coastal settlements (Adgar 1999). In this way, mangroves are the only vegetation in coastal habitat management worldwide.

Watershed Management

In the hydrological processes of groundwater recharge, water infiltration, evapotranspiration, rainfall, etc., ecosystems play an influential role. In the conditions of deficiency or floods, the influential role of ecosystems may also be changed. In Indonesia, elevated base flow in forested water sheds lowers the effects of dryness on subsequent farming (Pattanayak and Kramer 2001).

Ecosystem Services for Wetlands

Wetlands are highly productive ecosystems. Nevertheless, wetlands' vast and imperceptible environmental services still need to be recognized and assessed. Wetlands offer a wide range of Provisioning, Regulating, Cultural, and Supporting Services.

Provisioning Services

Incorporate the advantages and resources of wetlands including freshwater, sustenance, textiles, energy supplies, genetic materials, biochemical substances, natural remedies, and pharmaceuticals.

Regulating Services

Incorporate the advantages of regulating ecosystem services such as water regulation, erosion regulation, water purification, waste regulation, and climate and natural hazards (e.g., floods, storms, and droughts).

Cultural Services

Incorporate the intangible advantages individuals derive from wetlands, such as spiritual elevation, intellectual growth, contemplation, leisure activities, and aesthetic encounters. Additionally, the potential for ecotourism should be considered, as well as the cultural diversity, knowledge systems, educational values, social connections, attachment to a specific location, artistic legacy, and the potential for ecotourism.

Supporting Services

Incorporate the essential components required for the generation of all further ecosystem services. The positive effects of these ecosystem services are indirect and occur over a long period, such as primary production, water cycling, and nutrient cycling (Fig. 1.3).

Fig. 1.3 A view wetland showing its great ecological significance and services. Photo captured by Moharana Choudhury

Regulating Services at Zonal and Continental Level

Biodiversity plays a role in rainfall reprocessing and producing aerial water spray flows at regional and continental levels. According to Ellison et al. (2012), evapotranspiration by the forests lowers the total water streams in a watershed. However, it promotes the backpump of water into the atmosphere, which results in plentiful rainfall at the regional level. Forests are also responsible for maintaining the level of moisture in the atmosphere. It is, however, established that forests are equally important in managing the adverse effects of droughts due to global climate change.

Outcomes of the Study

Many projects and programs contribute equally to effectively implementing global ecosystem services for climate resilience. It has been estimated that by 2050, around two-thirds of humanity will be urban dwellers, and then cities will be the main component of climate variation and alteration the world over. From the above study, biodiversity will certainly be a salient factor in dealing with variations in weather. It is now successfully established that climate resilience is a universal problem requiring multijurisdictional, multilateral, and multinational governance. In the years to come, there will be great threats before human civilization, such as decarbonization, stability in the concentration of global greenhouse gases, and the temperature of the earth at its peak. There will be only one solution for the gradual cooling down of the temperature of the earth's surface: Restoration and protection of the ecosystems.

As we know, climate change is unstoppable. Not later in the coming decades will societies and ecosystems become substantial, or committed actions will have been taken to control global warming through NbS (Nature Based Solutions). Ecosystem restoration and management will be beneficial for long periods in climate change mitigation and community adaptation. Conjugative application of these practices, along with a reduction in emissions from fossil fuels, will always be recommended for better results.

Quantitative and Qualitative Analysis

Ecosystem services contribute to climate mitigation as they can minimize and remove the concentration of carbon particles from the atmosphere. It has been estimated that terrestrial ecosystems can absorb around 3 billion tons of environmental carbon each year (Pg/Yr) through net development, which represents 30% of anthropogenic CO_2 outflows in which timberland environments, especially tropical forests assume a vital part in carbon sequestration, while the deforestation of tropical forests results into 0.8–2.8 (Pg/Yr) carbon emissions (Baccini et al. 2012) which are comparable to 6–17% of worldwide and anthropogenic atmospheric CO_2 emanations (Van der Werf et al. 2009). In this manner, conventional environment management systems can effectively add to climate change mitigation. Conversion of old non-forested land to new woodland (afforestation) and new non-forested land to new backwoods (reforestation) directly reduces fossil fuel emissions. Agricultural management practices, including soil preservation and presentation of trees in agroforestry service, can enhance carbon sequestration (Uprety et al. 2012) (Fig. 1.4).

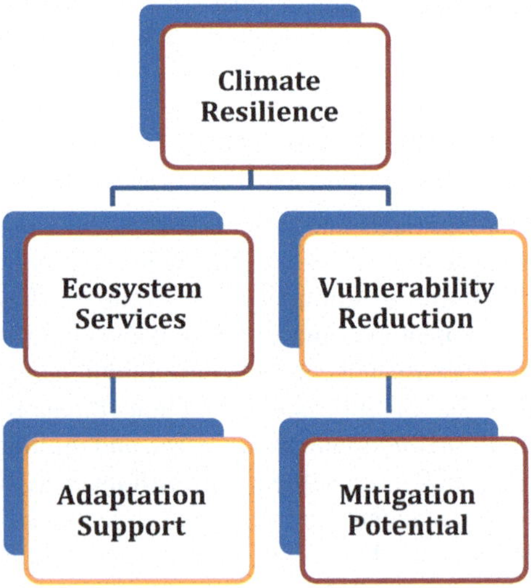

Fig. 1.4 Theoretical framework: climate resilience, ESs, and vulnerability reduction

To summarise, non-sustainable planning and infrastructural development are independent of technical or design instruments. Strategic Environmental Assessment (SEA) and Environmental Impact Assessment (EIA) practices have been developed to assess the potential impact on the environment. Similarly, the International Organization for Standardization (ISO) publishes ISO-14090: Adaptation to Climate Change: "Principles, requirements and guidelines" for the successful implementation of various ecosystem services in promoting climate resilience worldwide.

Conclusion

In this research, the researchers have explored the fundamental nexus between ecosystem services and climate resilience, aiming to provide a comprehensive theoretical framework for understanding their interconnections. Ecosystem services, encompassing provisioning, regulating, cultural, and supporting services, are crucial in bolstering the resilience of communities and ecosystems to climate change impacts. Through our literature synthesis, we identified key themes highlighting how various ecosystem services enhance adaptive capacity, reduce vulnerability, and improve overall resilience. However, gaps still need to be in our understanding, particularly in quantifying the economic value of ecosystem services and scaling up successful interventions across diverse socio-ecological contexts. Future research should address these gaps, leveraging interdisciplinary approaches to refine theoretical frameworks and inform policy and decision-making processes. Ultimately, integrating ecosystem services into climate resilience strategies offers a pathway toward building more adaptive, sustainable, resilient societies. By recognizing the inherent value of nature and harnessing its capacity to provide essential services, we can navigate the challenges posed by climate change more effectively and safeguard the well-being of current and future generations.

References

Adgar WN (1999) Social vulnerability to climate change and extremes in coastal Vietnam. World Dev 27:249–269

Baccini A, Goetz SJ, Walker WS et al (2012) Estimate carbon dioxide emissions from tropical deforestation improved by carbon-density maps. Nat Clim Change 2:182–185

Bele M, Somorin O, Sonwa D, Nkem J, Locatelli B (2011) Forests and climate change adaptation policies in Cameroon. Mitig Adapt Strat Glob Change 16:369–385

Bennett EM, Peterson GD, Gordon LJ (2009) Understanding relationships among multiple ecosystem services. Ecol Lett 12:1394–1404

Birkhofer K, Diehl E, Andersson J, Ekroos J, Fruh-Muller A, Machnikowski F, Mader VL, Nilsson L, Sasaki K, Rundloff M (2015) Ecosystem services-current challenges and opportunities for ecological research. Front Ecol Evol 2:87

Carter Berry Z, Jones KW, Gomez Aguilar LR, Congalton RG, Holwerda F, Kolka R, Looker N, Cogalton RG, Holwerda F, Lopez Ramirez SM, Manson R, Mayor A et al (2020) Evaluation of ecosystem service trade-offs along a land-use intensification gradient in Central Veracruz, Mexico. Ecosyst Serv 45:101181

Choudhury M, Sharma A, Singh P, Kumar D (2021) Impact of climate change on wetlands, concerning Son Beel, the largest wetland of North East, India. In: Global climate change. Elsevier, pp 393–414. https://doi.org/10.1016/B978-0-12-822928-6.00006-X

Cord AF, Bartkowski B, Bechmann M, Dittrich A, Hermans-Neumnn K, Kaim A, Lienhoop N, Locher-Krause K, Priess J, Schroter-Schlaack C et al (2017) Towards systematic analysis of ecosystem services trade-offs and synergies: main concepts, methods and road ahead. Ecosyst Serv 28:264–272

Costanza R, de Groot R, Braat L, Kubiszewski I, Fioramonti l, Sutton P, Farber S, Grasso M (2017) Twenty years of ecosystem services: how far have we come and how far do we still need to go? Ecosyst Serv 28:1–16

Dade MC, Mitchell MGE, McAlpine CA, Rhodes JR (2019) Assessing ecosystem service trade-offs and synergies: the need for more mechanistic approach. Ambio 48:1116–1128

Daily GC, Polasky S, Goldstein J, Kareiva PM, Mooney HA, Pejchar L, Ricketts TH, Salzman J, Shallenberger R (2009) Ecosystem services in decision making: time to deliver-enhanced reader. Front Ecol Environ 17:21–28

Das S, Vincent JR (2009) Mangroves protected villages and reduced death toll during Indian super cyclone. Proc Natl Acad Sci U S A 106:7357–7360

Delibas M, Tezer A, Kuzniecow Bacchin T (2021) Towards embedding soil ecosystem services in spatial planning. Cities 113:103150

Djoudi H, Brockhaus M, Locatelli B (2013) Once there was a lake: vulnerability to environmental changes in Northern Mali. Reg Environ Change 13:493–508

Doswald N, Munroe R, Roe D et al (2014) Effectiveness of ecosystem based approaches for adaptation: review of the evidence base. Clim Dev 6(2):185–201

Ellison D, Futter MN, Bishop K (2012) On the forest water yield debate: from demand-to supply-side thinking. Glob Change Biol 18:806–820

Fisher M, Chaudhary M, McCusker B (2010) Do forests help rural households adapt to climate variability? Evidence from Southern Malawi. World Dev 38:1241–1250

Garrity DP, Akinnifesi FK, Ajayi OC et al (2010) Evergreen agriculture: a robust approach to sustainable food security in Africa. Food Secur 2:197–214

Gill SE, Handley JF, Ennos AR, Pauleit S (2007) Adapting cities for climate change: the role of the green infrastructure. Built Environ 33:115–133

Gogoi N, Sarma A, Choudhury M, Samanta P, Faizan M, Sadhak S (2024a) Soil heavy metal pollution and ecological risk assessment in disturbed and undisturbed soil of Morigaon, Assam. Discover Environ 2(1):13. https://doi.org/10.1007/s44274-024-00039-z

Gogoi N, Choudhury M, Ul Hasan MS, Changmai B, Baruah D, Samanta P (2024) Soil organic carbon pool in diverse land utilization patterns in North-East India: an implication for carbon sequestration. Environ Dev Sustain, 1–27. https://doi.org/10.1007/s10668-024-05037-y

Gorman CE, Torsney A, Gaughran A, McKeon CM, Farrell CA, White C, Donohue I, Stout JC, Buckley YM (2023) Reconciling climate action with the need for biodiversity protection, restoration and rehabilitation. https://doi.org/10.1016/j.scitotenv.2022.159316

Han HQ, Liu Y, Gao HJ, Zhang YJ, Wang Z, Chen XQ (2020) Trade-offs and synergies between ecosystem services: a comparison of the karst and Non-Karst area. J Mt Sci 17:1221–1234

Iftekhar MS, Takama T (2008) Perception of biodiversity, environmental services, and conservation of planted mangroves: a case study on Nijhum Dwip Island, Bangladesh. Wetlands Ecol Manage 16(2):119–137

Kumar D, Choudhury M, Rathore A (2020) Valuation of ecosystem services & benefits of Son Beel wetland in Assam, India: a case study of natural solutions to climate change & water: the worldwide online climate change conference. Hamburg University, Germany

Kumar D, Choudhury M (2021) Recognizing economic values of wetland ecosystem services: a study of emerging role of monetary evaluation of Chandubi ecosystem and biodiversity. In:

Wetlands conservation: current challenges and future strategies, pp 87–110. https://doi.org/10.1002/9781119692621.ch5

Kumar P, Thakur PK, Bansod BKS, Debnath SK (2018) Groundwater: a regional resource and regional governance. Environ Dev Sustain 20:1133–1151

Kumar A, Kumar M, Pandey R, Yu Z, Pintod M (2021) Forest soil nutrient stocks along with an altitudinal range of Uttarakhand Himalayas: an aid to nature based climate solutions. CATENA 207:105667

Lee H, Lautenbach SA (2016) A quantitative review of relationships between ecosystem services. Ecol Indic 66:340–351

Li J, Zhang C, Zhu S (2021) Relative contributions of climate and land use change to ecosystem services in Arid Inland Basins. J Clean Prod 298:126844

Lin BB (2010) The role of agro forestry in reducing water loss through soil evaporation and crop transpiration in Coffee agro ecosystems. Agric for Meteorol 150:510–518

Liswanti N, Sheil D, Basuki I, Padmanaba M, Mulcahy G (2011) Falling back on forests: how forest-dwelling people cope with catastrophe in a changing landscape. Int for Rev 13:442–455

McSweeney K (2005) Natural insurance, forest access, and compounded misfortune: forest resources in smallholder coping strategies before and after Hurricane Mitch, Eastern Honduras. World Dev 33:1453–1471

Millennium Ecosystem Assessment (2005) Ecosystems and Human well-being: the assessment series (four volumes and summary). Island Press, Washington, DC

Millennium Ecosystem Assessment (MEA) (2005) Ecosystems and human well being: synthesis. Island Press, Washington, USA

Nelson E, Sander H, Hawthorne P, Conte M, Ennaanay D, Wolny S, Manson S, Polasky S (2010). Projecting global land-use changes and its effect on ecosystem service provision and biodiversity with simple models. PLOS ONE 5:e14327

Pattanayak SK, Kramer R (2001) Worth of watersheds: a producer surplus approach for valuing drought mitigation in Eastern Indonesia. Environ Dev Econ 6:123–146

Peters MK, Hemp A, Appelhans T, Becker JN, Behler C, Classen A, Detsch F, Ensslin A, Ferger SW, Frederiksen SB et al (2019) Climate-land use interactions shape tropical mountain biodiversity and ecosystem functions. Nature 568:88–92

Rai P, Vineeta, Shukla G, Manohar AK, Bhat JA, Kumar A, Kumar M (2021) Carbon storage of single tree and mixed tree dominant species stands in a reserve forest-case study of the Eastern Sub-Himalayan region of India. Land 10:435

Roberts D, Bonn R, Diederichs N et al (2012) Exploring ecosystems-based adaptation in Durban, South Africa: "learning-by—doing" at the local government Coal face. Environ Urban 24:167–195

Robledo C, Fischler M, Patino A (2004) Increasing the resilience of hillside communities in Bolivia: has vulnerability to climate change been reduced as a result of previous sustainable development corporation? Mt Res Dev 24:14–18

Sattler C, Loft L, Mann C, Meyer C (2018) Methods in ecosystem services governance analysis: an introduction. Ecosyst Serv 34:155–168

Sen S, Guchhait SK (2021) Urban green space in India: perception of cultural ecosystem services and psychology of situatedness and connectedness. Ecol Indic 123:107338

Shukla AK, Ojha CSP, Garg RD, Shukla S, Pal L (2020) Influence of spatial urbanization on hydrological components of the Upper Ganga River Basin, India. J Hazard. Toxic Radioact Waste 24:04020028

Shukla AK, Ojha CSP, Mijic A, Buytaert W, Pathak S, Garg RD, Shukla S (2018) Population growth, land use and land cover transformations and water quality Nexus in the Upper Ganga River Basin. Hydrol Earth Syst Sci 22:4745–4770

Smith P, Keesstra SD, Silver WL, Adhya TK (2021) The role of soils in delivering nature's contributions to people. Philos Trans R Soc B 376(1834):20200169. https://doi.org/10.1098/rstb.2020.0169

Uprety DC, Dhar S, Hongmin D et al (2012). Technologies for climate change mitigation: agriculture sector. UNEP Riso Centre of Energy, Climate and Sustainable Development, Copenhagen

Vander Werf GR, Morton DC, De Fries RS et al (2009) CO_2 emissions from forest loss. Nat Geosci 2:737–738

Verchot LV, Van Noordwijk M, Kandji S et al (2007) Climate change: linking adaptation and mitigation through agro forestry. Mitig Adapt Strat Glob Change 12:901–918

Walton MEM, Samonte-Tan GPB, Primavera JH, Edwards-Jones G, LeVay L (2006) Are mangroves worth replanting? The direct economic benefits of a community-based reforestation project. Environ Conserv 33:335–343

Wang LJ, Ma S, Zhao YG, Zhang JC (2021) Ecological restoration projects did not increase the value of all ecosystem services in Northern China. For Ecol Manag 495:119340

Wu C, Chen B, Huang X, Dennis Wei YH (2020) Effect of land-use change and optimization on the ecosystem service values of Jiangsu province, China. Ecol Indic 117:106507

Xia M, Jia K, Wang X, Bai X, Li C, Zhao W, Hu X, Cherubini F (2021) A framework for regional ecosystem authenticity evaluation—a case study on the Qinghai-Tibet plateau of China. Glob Ecol Conserv 31:e01849

Xue L, Zhu B, Wu Y, Wei G, Liaos S, Liao S, Yang C, Wang J, Zhang H, Ren L, Han Q (2012) Dynamic protection of ecological risk in the Manas river basin based upon terrain gradients. Sci Total Environ 635:283–293

Zhang L, Cheng L, Chiew F, Fu B (2018) Understanding the impacts of climate and land change on water yields. Curr Opin Environ Sustain 33:167–174

Chapter 2
A Comprehensive Insight on Ecosystem Services and Climate Resilience

Murad Muhammad, Shumaila Batool, Muqadas Batool, Sana Ullah, Mohammad Reza Morshedloo, Abdul Wahab, Mukul Machhindra Barwant, and Shahid Ullah

Abstract Ecosystem services play a vital role in improving climate resilience by providing essential functions that support the stability and adaptation of ecosystems in the face of changing climate. This interwoven relationship emphasizes the crucial importance of natural systems in improving resilience and maintaining human well-being in a shifting climate environment. Ecosystem services include many advantages, such as soil fertility maintenance, pollination, water purification, and natural

M. Muhammad (✉)
Xinjiang Key Laboratory of Biodiversity Conservation and Application in Arid Lands, Xinjiang Institute of Ecology and Geography, Chinese Academy of Sciences, Urumqi, People's Republic of China
e-mail: Muradbotany1@uop.edu.pk

M. Muhammad · A. Wahab
University of Chinese Academy of Sciences, Beijing 100049, People's Republic of China

M. Muhammad
State Key Laboratory of Desert and Oasis Ecology, Key Laboratory of Ecological Safety and Sustainable Development in Arid Lands, Xinjiang Institute of Ecology and Geography, Chinese Academy of Sciences, Urumqi 830011, People's Republic of China

S. Batool
Department of Zoology, Wildlife and Fisheries, University of Agriculture, Faisalabad 38000, Pakistan

M. Batool
Department of Botany, (Govt Graduate College Layyah) Bahauddin Zakariya University, Multan, Pakistan

S. Ullah
Department of Agricultural Leadership, Education & Communication College of Agricultural & Environmental Sciences, University of Georgia, Athens, GA 30602, USA

M. R. Morshedloo
Department of Horticultural Sciences, Faculty of Agriculture, University of Maragheh, Maragheh, Iran

A. Wahab
Shanghai Center for Plant Stress Biology, CAS Center for Excellence in Molecular Plant Sciences, Chinese Academy of Sciences, Shanghai 200032, People's Republic of China

M. Choudhury et al. (eds.), *Sustainable Synergy: Harnessing Ecosystems for Climate Resilience*, Climate Change Management,
https://doi.org/10.1007/978-3-031-77957-2_2

hazard mitigation. These services are essential to climate resilience strategies as they enhance ecosystem health and performance. Ecosystem services improve ecosystems' ability to tolerate environmental stressors, providing vital resources for human societies like food security, clean water, and protection from natural disasters. It is crucial to focus on conserving and sustainably managing ecosystem services to promote climate resilience, enhance adaptive capacity, and decrease vulnerability to climate-related issues. Acknowledging the inherent worth of ecosystem services and incorporating them into climate adaptation strategies can improve the resilience and sustainability of communities, making them more capable of dealing with the challenges of a changing climate. By embracing and protecting ecosystem services, more efficient climate resilience solutions can be developed to benefit ecosystems and human well-being in the long run. By responsibly preserving and managing ecosystem services, we can promote a more resilient and sustainable future for everyone.

Keywords Climate resilience · Ecosystem services · Biodiversity · Resilience

Introduction

The importance of ecosystem services in adapting to and mitigating the effects of climate change is critical (Munang et al. 2013). The goal of mitigation is to either lessen the production of greenhouse gases or make natural systems better able to absorb them. Adaptation aims to make changes to both human and non-human systems to reduce the impact of climate change or take advantage of its potential benefits (Waheed et al. 2023). Mitigation and adaptation strategies prioritize specific sectors and areas based on their justifications. While adaptation seeks to lessen the impact on already vulnerable ecosystems, people, and activities, mitigation efforts mainly aim at reducing emissions from large-scale emitters or powerful sinks (Bhatt 2017). Ecosystems play a crucial role in mitigating the impact of climate change due to their exceptional capacity to absorb and retain carbon dioxide (Gogoi et al. 2024). Ecosystems play a vital part in the adaptation process by providing services that humans may utilize to deal with the current and future hazards caused by climate change. While ecosystem services play a vital role in mitigating the effects of global warming, they are nonetheless susceptible to the consequences of climate change (Kumar and Choudhury 2021). Understanding the mutual reliance of ecosystem services and climate change is crucial for the efficacy of ecosystem-based initiatives aimed at combating climate change (Chong 2014). We can maximize the ecosystem's

M. M. Barwant
Department of Botany, Sanjivani Arts Commerce and Science College Kopargaon, Kopargaon, Maharashtra, India

S. Ullah
Department of Botany, University of Peshawar, Peshawar, Pakistan

support for adaptation and mitigation efforts—also called "ecosystem-based adaptation and mitigation"—through ecosystem management. Adaptation strategies, also called "ecosystem services adaptation," are necessary to lessen the impact of climate change and keep ecosystems functioning generally because of the expected effects on ecosystems and the services they provide (Skalová et al. 2022).

Optimal management of land, water, soil, and genetic resources—among other natural resources—is essential to a system's resilience, defined as its capacity to recover from setbacks (Cosgrove and Loucks 2015). Robust and resilient ecosystems can boost societal and economic prosperity by ensuring a consistent supply of ecosystem services that enhance people's quality of life and by providing a solid foundation for various sectors (including fishing, recreation, and tourism), Protecting against storm surges and managing floods are two examples of how resilient ecosystems can mitigate disturbances that could lower income or quality of life (Hobbie and Grimm 2020).

The welfare of humans and the health of the environment are closely related. When managing diseases, ecosystems are essential in directly improving human health disease vectors like ticks and mosquitoes can be better managed in an ecosystem rich in biodiversity and home to many animals and insects (Shah et al. 2020; Rayan et al. 2021). We might expect climate-induced changes in the distribution of productive ecosystems and the accessibility of resources like food, water, and energy to affect human health negatively (Muhammad et al. 2024). These shifts will profoundly affect the spread of infectious diseases, community nutrition, and the layout of human settlements (Bereitschaft and Scheller 2020).

It is critical to recognize the existing inequalities in the effects of climate change to address the need for social justice and equity in the context of climate resilience. Extreme weather events exacerbate existing inequities by disproportionately impacting already marginalized and economically disadvantaged populations (Smiley et al. 2022). Therefore, these communities' needs and perspectives must be considered in any discussion of adaptation and mitigation solutions to handle their specific problems effectively. This includes acknowledging their weaknesses and giving them the tools to be involved in decision-making. To better tailor resilience initiatives to satisfy the requirements of all stakeholders, it is essential to adopt an inclusive strategy that includes varied viewpoints and experiences. More resilient and united communities, where the costs and advantages of climate action are shared relatively, result from incorporating social justice and equity concepts into resilience policies and practices (Wenta et al. 2019). In the end, we can strengthen community resilience and protect fundamental human rights and principles of fairness against the effects of climate change by making social justice and equity the top priorities in climate resilience programs.

Understanding Ecosystem Services

Ecosystem Services and Climate Resilience

Ecosystem services refer to the natural products, situations, or activities that enhance human or societal well-being. Individuals can benefit from ecosystem services as a final product or as a constituent in other goods. An illustration of an ecosystem service is the pollination of crops by bees and other organisms, which contributes to the augmentation of food production (Birkhofer et al. 2015). An additional example is the utilization of riparian buffers and wetlands to mitigate floods in residential areas. The Millennium Ecosystem Assessment (MA), a comprehensive UN-sponsored initiative, identified four primary categories of ecosystem services: provisioning, regulatory, cultural, and sustaining services (Pu et al. 2023). These services were examined to understand the impact of human activities on ecosystems and human well-being. When we talk about climate resilience, we mean the ability to foresee, prepare for, and respond to potentially harmful climate-related events, patterns, or disruptions. To make communities more resistant to the effects of climate change, we need to figure out how this phenomenon will alter current climate-related risks or create new ones and how to handle them better (Cianconi et al. 2020). Ecosystem resilience to climate change can be better understood with a holistic paradigm that considers community engagement, indicator tracking, system adaptability, the impact of climate change on ecosystem services, and expansion across spatial and temporal scales (Paschen and Ison 2014). Table 2.1 shows ecosystem resilience, which is the capacity of an ecosystem to maintain its functioning and adapt to new conditions, even when faced with extreme events like climate change. Subterranean microbial communities, like arbuscular mycorrhizal fungi, play an essential role in ecosystem resilience and adaptation by influencing plant mineral uptake and productivity; this role should be included in this paradigm shift (Muhammad et al. 2023).

Adaptation and Mitigation Strategies

Adapting to climate change involves altering one's conduct in reaction to the ongoing changes occurring on Earth (Pelling et al. 2015). This can only be accomplished if people are aware of the harm climate change will do and do something about it, whether preparing for it, adapting to it, or mitigating it. Another possible meaning of the term is capitalizing on opportunities climate change brings. Preparation for a shifting climate should have been implemented well in advance by establishing adaptation strategies. The present now is the second most suitable period. Both the present and future climatic extremes will ensure you implement preventive measures. Enhanced security for citizens, property, and infrastructure can be attained by ensuring that cities and regions are prepared to adjust to the impacts of climate change (Balogun et al. 2020). The adverse influences and economic problems caused

Table 2.1 The relationship between climate resilience and ecosystem services

Climate resilience	Description	Ecosystem services impacted	References
Increased extreme events	The consequences of climate change include the occurrence of more frequent and severe weather events, such as floods, storms, and droughts, which have the potential to disturb ecosystems and reduce their capacity to offer services	Regulating services like flood management, erosion control, and water purification can be compromised, impacting water availability, soil stability, and disaster risk reduction	Kiedrzyńska et al. (2015)
Temperature changes	Rising temperatures can alter habitat suitability for species, affecting biodiversity and ecosystem functions	Changes in species distribution and crop yields may affect providing services like food production, while regulating services like climate regulation may be compromised	Morán-Ordóñez et al. (2021)
Altered precipitation	Alterations in precipitation patterns, such as alterations in the intensity of rainfall and their timing, can influence water availability and quality, affecting various ecosystem services	Provisioning services like water supply for drinking and agriculture, regulating services like water purification, and supporting services like the water cycle can be affected	Chang and Bonnette (2016)
Sea level rise	The rise in sea levels can result in saltwater intrusion, coastal erosion, and the loss of habitat, all of which have the potential to impact ecosystems and the services that they contribute, particularly in coastal regions	Regulating services like erosion control and coastal protection, provisioning services like seafood production, and cultural services like tourism and recreation can be impacted	Wahab et al. (2024)
Biodiversity loss	Climate change can exacerbate biodiversity loss through habitat destruction, extinction, and altered ecological interactions, reducing ecosystems' resilience and capacity to provide services	All ecosystem services can be impacted as biodiversity loss undermines the functioning and stability of ecosystems, affecting their ability to provide provisioning, regulating, a	Isbell et al. (2015)

by climate change can be reduced by enhancing resilience and improving the ability to adapt. Collaboration between urban areas and regions can lead to the attainment of sustainability goals and enhance the well-being of both residents and tourists (Dwyer 2023). Without proper coordination and preparation, the likelihood of successful adaptation is highly improbable. An integrated approach to adaptation combines building resilience to climate change with reducing a city's or region's contribution to the problem, known as mitigation (Lin et al. 2021). This approach also offers various co-benefits, such as protecting citizens' health, providing recreational opportunities, and preserving natural habitats and biodiversity, as shown in Table 2.2. Combining these components makes it possible to maximize advantages while simultaneously minimizing expenses. Reducing greenhouse gas emissions and increasing carbon sequestration rates should be the primary goals of sustainable adaptation strategies, such as ecosystem-based adaptation. These strategies should contribute to mitigating climate change (Naumann et al. 2011).

Table 2.2 Outlining adaptation and mitigation strategies for climate resilience

Strategy	Adaptation	Mitigation	References
Infrastructure upgrade	Reinforce buildings and roads for extreme weather	Develop green infrastructure, renewable energy	Auld and MacIver (2005)
Ecosystem conservation	Protect and restore habitats to buffer against climate impacts	Preserve forests and wetlands to absorb CO_2	Mitra et al. (2005)
Water management	Implement water conservation measures, build resilient water systems	Reduce water usage, invest in water-efficient technologies	Maggioni (2015)
Agriculture	Shift to climate-resilient crops, adopt new farming techniques	Promote sustainable farming practices, reduce emissions from agriculture	Muhammad et al. (2023)
Urban planning	Design cities for resilience, enhance drainage systems	Encourage public transportation, reduce urban sprawl	Zhao (2010)
Education and awareness	Educate communities on climate risks, promote adaptive behaviours	Raise awareness about carbon footprint, advocate for renewable energy	Cotton et al. (2015)

Challenges and Limitations

Data and Measurement Issues

Data availability and quality pose significant challenges to integrating ecosystem services into climate policy. Measuring ecological services such as carbon sequestration, water control, and biodiversity is complex and resource-intensive. Moreover, data limitations, particularly in developing nations, may hinder the comprehensive representation of the benefits of ecosystem services. Additionally, the intricate interaction between biotic and abiotic elements further complicates the assessment of ecosystem resilience to climate change.

Policy and Implementation Gaps

Another major obstacle in incorporating ecosystem services into climate policy is the difficulty involved. While ecosystem services are crucial, lawmakers need help to fully understand their value and determine how to integrate them into existing legislative frameworks. This challenge is further compounded by the fact that managing ecosystem services often requires collaboration across multiple sectors and government levels, necessitating overcoming additional institutional and governance hurdles. The presence of unclear legal frameworks, inadequate incentives, and enforcement mechanisms may fail policies to preserve and restore ecosystems.

Socio-economic and Equity Considerations

Incorporating environmental services into climate policy also brings significant socio-economic and equity considerations. The distributional effects of ecosystem-based techniques can range among various communities, and specific vulnerable populations, such as indigenous communities and low-income groups, may experience a disproportionate influence. It is essential to fairly allocate the advantages and disadvantages of ecosystem-based climate solutions to tackle social justice issues and foster inclusive climate action (Kumar et al. 2020).

The difficulties and restrictions in integrating ecosystem services into climate policy encompass data and assessment problems, policy and implementation gaps, and socio-economic and equity factors. To overcome these obstacles, a comprehensive strategy is necessary. This strategy should focus on expanding data collecting and monitoring, improving policy and governance frameworks, and prioritizing the needs and views of all stakeholders, especially those most vulnerable.

Future Perspective and Recommendations

Looking into the future, addressing ecosystem services and climate resilience will become increasingly crucial as climate change impacts intensify. Here are some future perspectives and recommendations:

Integrated Planning and Management: Future efforts should focus on integrating ecosystem services and climate resilience considerations into all planning and management levels, from local to global scales. This includes land-use planning, conservation strategies, and infrastructure development (Sajid et al. 2020).

Nature-Based Solutions: It is essential to emphasize the significance of nature-based solutions for climate resilience. Investing in and restoring natural ecosystems like mangroves, woodlands, and wetland areas can bring several benefits, including preventing flooding, storing carbon, and preserving habitat (Beck et al. 2022).

Green Infrastructure: Promote the development of green infrastructure projects that enhance ecosystem services while providing climate adaptation benefits (Lovell and Taylor 2013). Examples of urban green spaces that reduce floods and lessen the effects of urban heat islands include green roofs, permeable pavements, and urban green spaces of various types.

Community Engagement and Equity: Ensure efforts to enhance ecosystem services and climate resilience are inclusive and equitable, particularly in vulnerable communities. Engage local communities in decision-making processes, incorporate traditional knowledge, and prioritise actions that address the needs of marginalised groups (Lynam et al. 2007).

Policy Support and Funding: Advocate for supportive policies and increased funding for ecosystem restoration, conservation, and climate adaptation initiatives. Governments and international organisations should prioritise investments in nature-based solutions and provide financial incentives for sustainable land management practices (Hallstein and Iseman 2021).

Research and Innovation: Encourage research and innovation better to understand the complex interactions between ecosystems and climate resilience. This includes developing predictive models, monitoring systems, and new technologies to assess ecosystem health, anticipate climate impacts, and inform adaptive management strategies.

Education and Awareness: Raising public awareness and educating people about the significance of ecosystem services and climate resilience should be done. Encourage a more in-depth comprehension of the interrelationships that exist between the natural world, the climate, and the well-being of humans, and give individuals and communities the ability to take action to preserve and repair ecosystems (Chapin et al. 2011).

Conclusion

Anticipated climate change impacts on ecosystem services are predominantly detrimental. There is widespread international concern regarding the depletion of ecosystem service supplies or the deterioration of the relationship between future demand and supply capacity. Conversely, the provision of energy, food, and fiber has witnessed substantial growth in both supply and demand, frequently to the detriment of the services above. Among the many contributing factors are climate change and mitigation initiatives. Increasing demand for services, population expansion, resource consumption per capita, ecosystem degradation, inadequate resource management, biodiversity loss, and chemical excess are all examples of non-climatic drivers. Climate change is anticipated to amplify these tendencies and restrict the expansion of food and fiber-based services that have been prevalent for the last century. Mismatched phenological responses, adaptation to climate shifts, and exceeding hard-wired limits such as grain formation or mammalian body temperature contribute to ecosystem disruption. Climate change will severely affect some ecosystem services more severely than others; for example, temperate regions will be less affected than hotter, drier, and more polar areas. Over the next hundred years, this pattern is expected to gain even more prominence. The region's most notable impacts are inadequate human welfare, population growth, and negligible historical impact on global climate change.

Consequently, the impacts are not uniformly dispersed. The disparity between local and global effects may result in conflict and inequality. Local food crop demands and the global demand for climate change mitigation services, including carbon sequestration and bioenergy production, may collide. Certain services (e.g., food and wood) are consumed more per capita in developed nations, whereas consumption growth is more rapid in emergent nations, especially in regions susceptible to climate change.

References

Auld H, MacIver D (2005) Cities and communities: the changing climate and increasing vulnerability of infrastructure. In: Adaptation and impacts research group, meteorological service of Canada ….

Balogun A-L, Marks D, Sharma R, Shekhar H, Balmes C, Maheng D, Arshad A, Salehi P (2020) Assessing the potentials of digitalization as a tool for climate change adaptation and sustainable development in urban centres. Sustain Cities Soc 53:101888

Beck MW, Heck N, Narayan S, Menéndez P, Reguero BG, Bitterwolf S, Torres-Ortega S, Lange G-M, Pfliegner K, McNulty VP (2022) Return on investment for mangrove and reef flood protection. Ecosyst Serv 56:101440

Bereitschaft B, Scheller D (2020) How might the COVID-19 pandemic affect 21st century urban design, planning, and development? Urban Sci 4(4):56

Bhatt RP (2017) Hydropower development in Nepal—climate change, impacts and implications. Renew Hydropower Technol 1:75–98

Birkhofer K, Diehl E, Andersson J, Ekroos J, Früh-Müller A, Machnikowski F, Mader VL, Nilsson L, Sasaki K, Rundlöf M (2015) Ecosystem services—current challenges and opportunities for ecological research. Front Ecol Evol 2:87

Chang H, Bonnette MR (2016) Climate change and water-related ecosystem services: impacts of drought in California, USA. Ecosyst Health Sustain 2(12):e01254

Chapin FS III, Power ME, Pickett ST, Freitag A, Reynolds JA, Jackson RB, Lodge DM, Duke C, Collins SL, Power AG (2011) Earth Stewardship: science for action to sustain the human-earth system. Ecosphere 2(8):1–20

Chong J (2014) Ecosystem-based approaches to climate change adaptation: progress and challenges. Int Environ Agreements Polit Law Econ 14:391–405

Cianconi P, Betrò S, Janiri L (2020) The impact of climate change on mental health: a systematic descriptive review. Front Psych 11:490206

Cosgrove WJ, Loucks DP (2015) Water management: current and future challenges and research directions. Water Resour Res 51(6):4823–4839

Cotton DR, Miller W, Winter J, Bailey I, Sterling S (2015) Developing student's energy literacy in higher education. Int J Sustain High Educ 16(4):456–473

Dwyer L (2023) Resident well-being and sustainable tourism development: the 'capitals approach.' J Sustain Tour 31(9):2119–2135

Gogoi N, Choudhury M, Ul Hasan MS, Changmai B, Baruah D, Samanta P (2024) Soil organic carbon pool in diverse land utilization patterns in North-East India: an implication for carbon sequestration. Environ Dev Sustain, 1–27. https://doi.org/10.1007/s10668-024-05037-y

Hallstein E, Iseman T (2021) Nature-based solutions in agriculture: project design for securing investment. Food & Agriculture Org.

Hobbie SE, Grimm NB (2020) Nature-based approaches to managing climate change impacts in cities. Philos Trans R Soc B 375(1794):20190124

Isbell F, Tilman D, Polasky S, Loreau M (2015) The biodiversity-dependent ecosystem service debt. Ecol Lett 18(2):119–134

Kiedrzyńska E, Kiedrzyński M, Zalewski M (2015) Sustainable floodplain management for flood prevention and water quality improvement. Nat Hazards 76:955–977

Kumar D, Choudhury M (2021) Recognizing economic values of wetland ecosystem services: a study of emerging role of monetary evaluation of Chandubi ecosystem and biodiversity. In: Wetlands conservation: current challenges and future strategies, pp 87–110. https://doi.org/10.1002/9781119692621.ch5

Kumar D, Choudhury M, Rathore A (2020) Valuation of ecosystem services & benefits of Son Beel wetland in Assam, India: a case study of natural solutions to climate change & water: the worldwide online climate change conference. Hamburg University, Germany

Lin BB, Ossola A, Alberti M, Andersson E, Bai X, Dobbs C, Elmqvist T, Evans KL, Frantzeskaki N, Fuller RA (2021) Integrating solutions to adapt cities for climate change. Lancet Planet Health 5(7):e479–e486

Lovell ST, Taylor JR (2013) Supplying urban ecosystem services through multifunctional green infrastructure in the United States. Landscape Ecol 28:1447–1463

Lynam T, De Jong W, Sheil D, Kusumanto T, Evans K (2007) A review of tools for incorporating community knowledge, preferences, and values into decision making in natural resources management. Ecol Soc 12(1)

Maggioni E (2015) Water demand management in times of drought: what matters for water conservation. Water Resour Res 51(1):125–139

Mitra S, Wassmann R, Vlek PL (2005) An appraisal of global wetland area and its organic carbon stock. Curr Sci 88(1):25–35

Muhammad M, Batool S, Hivare V, Li W-J, Waheed A, Sinha D (2024) Bioremediation techniques—classification, principles, advantages, limitations, and prospects. In: Microbiome-assisted bioremediation. Elsevier, pp 1–23

Muhammad M, Waheed A, Wahab A, Majeed M, Nazim M, Liu Y-H, Li L, Li W-j (2023) Soil salinity and drought tolerance: an evaluation of plant growth, productivity, microbial diversity, and amelioration strategies. Plant Stress, 100319

Munang R, Thiaw I, Alverson K, Liu J, Han Z (2013) The role of ecosystem services in climate change adaptation and disaster risk reduction. Curr Opin Environ Sustain 5(1):47–52

Naumann S, Anzaldua G, Berry P, Burch S, Davis M, Frelih-Larsen A, Gerdes H, Sanders M (2011) Assessment of the potential of ecosystem-based approaches to climate change adaptation and mitigation in Europe. Final report to the European Commission, DG Environment

Paschen J-A, Ison R (2014) Narrative research in climate change adaptation—exploring a complementary paradigm for research and governance. Res Policy 43(6):1083–1092

Pelling M, O'Brien K, Matyas D (2015) Adaptation and transformation. Clim Change 133:113–127

Pu X, Ding W, Ye W, Nan X, Lu R (2023) Ecosystem service research in protected areas: a systematic review of the literature on current practices and future prospects. Ecol Ind 154:110817

Rayan RA, Choudhury M, Deb M, Chakravorty A, Devi RM, Mehta J (2021) Climate change: impact on waterborne infectious diseases. In Water conservation in the era of global climate change. Elsevier, pp 213–228. https://doi.org/10.1016/B978-0-12-820200-5.00014-2

Shah S, Khan S, Sulaiman S, Muhammad M, Badshah L, Bussmann RW, Hussain W (2020) Quantitative study on medicinal plants traded in selected herbal markets of Khyber Pakhtunkhwa, Pakistan. Ethnobot Res Appl 20:1–36

Skalová B, Jaššo M, Husár M, Ondrejička V (2022) Cooperation of the city management with the public and private sector in relation to the implementation of climate mitigation and adaptation measures. In: AIP conference proceedings

Smiley KT, Noy I, Wehner MF, Frame D, Sampson CC, Wing OE (2022) Social inequalities in climate change-attributed impacts of Hurricane Harvey. Nat Commun 13(1):3418

Wahab A, Muhammad M, Ullah S, Abdi G, Shah GM, Zaman W, Ayaz A (2024) Agriculture and environmental management through nanotechnology: eco-friendly nanomaterial synthesis for soil-plant systems, food safety, and sustainability. Sci Total Environ, 171862

Waheed A, Haxim Y, Islam W, Ahmad M, Muhammad M, Alqahtani FM, Hashem M, Salih H, Zhang D (2023). Climate change reshaping plant-fungal interaction. Environ Res, 117282

Wenta J, McDonald J, McGee JS (2019) Enhancing resilience and justice in climate adaptation laws. Transnational Environ Law 8(1):89–118

Zhao P (2010) Sustainable urban expansion and transportation in a growing megacity: consequences of urban sprawl for mobility on the urban fringe of Beijing. Habitat Int 34(2):236–243

Chapter 3
Quantifying the Impact: Ecosystems Contribution to Climate Resilience

Prasann Kumar, Joginder Singh, and Anmoldeep Kaur

Abstract Ecosystems play a pivotal role in fostering climate resilience by regulating natural processes, mitigating climate-related risks, and enhancing the adaptive capacities of both human and ecological communities. This chapter explores the various ecosystem services that directly and indirectly support climate adaptation and mitigation efforts. By providing a comprehensive review of ecosystem-based approaches, the chapter delves into the mechanisms through which ecosystems, such as forests, wetlands, and coastal systems, buffer against extreme weather events, regulate carbon cycles, and sustain biodiversity, all of which are essential for long-term resilience. The chapter further introduces methodologies for quantifying the ecological, social, and economic impacts of ecosystem services in building resilience. It draws on case studies from around the globe to demonstrate how integrating ecosystem-based solutions into climate resilience strategies not only enhances environmental sustainability but also supports socio-economic development. The concluding sections discuss the challenges in quantifying these contributions, the tools available for assessment, and the need for more robust frameworks to measure and maximize ecosystem resilience in the face of climate change.

Keywords Ecosystem · Biodiversity · Conservations · Resilience · Carbon Sequestration

P. Kumar (✉)
Department of Agronomy, School of Agriculture, Lovely Professional University, Phagwara, Punjab 144411, India
e-mail: prasann0659@gmail.com

J. Singh
Department of Botany, Nagaland University, Nagaland, India

A. Kaur
Department of Agriculture, Royal Agricultural University, Cirencester GL7 6JS, Gloucestershire, UK

M. Choudhury et al. (eds.), *Sustainable Synergy: Harnessing Ecosystems for Climate Resilience*, Climate Change Management,
https://doi.org/10.1007/978-3-031-77957-2_3

Introduction

Climate resilience is the ability of individuals, communities, ecosystems, and systems to anticipate, endure, recover from, and potentially transform in response to climate change. Climate resilience means adapting and thriving in changing climates, including severe weather, temperature and precipitation changes, sea level rise, and other effects (Maebe et al. 2023; Lacambra et al. 2024; Mallick et al. 2024). Climate resilience is a complex concept that involves multiple dimensions. Preventive measures, such as early warning systems and risk assessments, reduce vulnerability and improve preparedness for climate-related events. Systems must be resilient to absorb and recover from disturbances without significant damage. Absorptive and adaptive capacities are closely related, focusing on systems' ability to adapt to changing weather conditions. Adaptability empowers individuals, communities, and institutions to address new challenges and seize opportunities. Transformative capacity, which involves significant changes in governance, institutions, values, and behaviours, is the pinnacle of climate resilience. Systemic changes and societal transformations are needed to build more fair and resilient societies. Ecosystems are crucial to climate resilience, as they help natural and human systems adapt to and mitigate climate change. Ecosystems enhance resilience in local communities and global ecological systems (Singh and Goyal 2023; Almulhim and Cobbinah 2024; Filippini et al. 2024; Li et al. 2024). Ecosystems are crucial in enhancing climate resilience and supporting life and climate change adaptation. They act as natural buffers, reducing the impact of extreme weather and climate hazards. Coastal ecosystems like mangroves and coral reefs protect coastal communities from storms and floods. Ecological systems also regulate Earth's climate through CO_2 sequestration and carbon storage, preserving carbon reserves and reducing vulnerability to depletion. Preserving and restoring these systems is essential for climate mitigation and resilience, providing critical resources and functions, strengthening human communities and livelihoods (Fig. 3.1).

Carbon Sequestration and Climate Regulation

Climate regulation and carbon sequestration are crucial for Earth's atmosphere and carbon cycle. Plants, trees, soils, and oceans store CO_2 to combat climate change. Sunlight-driven photosynthesis converts atmospheric CO_2 into organic compounds, sequestering soil carbon through organic matter and microbial activity (Gogoi et al. 2024a). Trees are the main carbon reservoirs in forests, storing carbon in biomass, roots, and soil organic matter (Wood and Dupras 2021; Ahmad et al. 2024; Sealey 2024). Oceans, marshes, swamps, and peatlands store carbon, regulating CO_2 and mitigating climate change. High-vegetation ecosystems store carbon through photosynthesis and underground biomass. Marine and coastal blue carbon environments

Fig. 3.1 Ecosystems' contribution to climate resilience. *Source* Created by authors

store more carbon, especially in coastal areas. Peatlands, wetlands, and grasslands regulate atmospheric CO_2 and store one-third of the world's soil carbon.

However, land conversion and peat fires contribute to climate change. Biodiversity and intact ecological systems help ecosystems sequester carbon and withstand global change (Azmi et al. 2023; Vilà-Cabrera et al. 2023; Qiao et al. 2024). Ecosystem carbon storage is crucial for climate regulation and global patterns, as it reduces atmospheric CO_2 emissions and slows global warming. Carbonaceous forests, wetlands, and grasslands protect against climate-related weather, flooding, and droughts. Ecological adaptation strategies help environmental systems and societies adapt to climate change. Depleting forests and peatlands releases large amounts of carbon into the atmosphere, worsening ecological systems and global warming. Global climate agreements acknowledge ecosystem carbon sequestration, emphasizing afforestation, reforestation, and sustainable land management. Forest carbon storage capacity depends on species composition and forest type. Reducing disturbances and land-use changes in forests can preserve carbon stocks and sequester carbon. Quantifying and monitoring forest carbon stocks helps determine carbon sequestration rates, storage changes, and climate mitigation strategies (Behboudian et al. 2023; Xie et al. 2023; Rhymes et al. 2024). Forest inventory and remote sensing estimate aboveground, belowground, and soil carbon stocks in forest ecosystems.

REDD+ and forest carbon offset projects promote forest preservation and rehabilitation. Carbon-sequestering Forest management can achieve economic, social, and environmental goals.

Forests provide timber, non-timber forest products, clean water, and recreation, helping rural communities thrives. Carbon sequestration, water regulation, soil conservation, and climate regulation are forest ecosystem services that improve human well-being and resilience. Sustainable forest management, biodiversity conservation, and deforestation and forest degradation reduction require international cooperation and partnerships. Forest conservation and restoration can meet global climate and sustainable development goals through public–private partnerships, voluntary commitments, and multi-stakeholder initiatives. For centuries, traditional knowledge and practices have helped Indigenous peoples and local communities sustainably manage forests and biodiversity (Kumar et al. 2022; Liu et al. 2023; Trégarot et al. 2024; Yuan et al. 2024). Rotational agriculture, agroforestry, and community-based resource management preserve biodiversity, sequester carbon, and strengthen ecosystems. Indigenous knowledge systems inform sustainable forest management and conservation by revealing forest ecology, species interactions, and ecological dynamics. Cooperation, empowerment, and inclusive forest management require respecting indigenous peoples' rights, knowledge, and governance systems. Forest science, technology, and innovation must improve to understand forest carbon dynamics and develop carbon sequestration and climate mitigation strategies (Singh et al. 2022).

Satellites and LiDAR can map worldwide forest cover, biomass, and carbon stock. Climate-smart forestry-assisted natural regeneration and BECCS may increase forest carbon sequestration and reduce greenhouse gas emissions. Wetlands and oceans store carbon and fight global warming, regulating Earth's climate (Gong et al. 2023; Mehmood et al. 2024; Meng et al. 2024; Parker et al. 2024). Marine ecosystems are including marshes, swamps, bogs, and mangroves, store carbon through photosynthesis and ocean sediment. They are essential carbon sequesters, storing organic matter. Ecosystem degradation could worsen climate change. Wetland restoration, sustainable coastal practices, and marine ecosystem protection are growing. Natural ecosystems absorb carbon dioxide, fighting climate change (Kumar et al. 2020). Restoring ecosystems, preserving ecological systems, and reducing human-caused emissions are crucial for a sustainable future (Fig. 3.2).

Water Resources Regulation

Water regulation is essential for the planet's vital resource's long-term survival and adaptability. This comprehensive analysis examines three critical components of water resource regulation: managing watersheds and climate resilience, regulating wetlands and floods, and examining ecosystems' role in water quality (Kumar and Choudhury 2021). This paper examines current practices, challenges, and future directions to show how effective regulation strategies can address the complex water

Fig. 3.2 Carbon sequestration and climate regulation. *Source* created by authors

resource issues of the twenty-first century. Water is essential to life, ecological systems, and human communities, but sustainable management is a global challenge (Sharma and Choudhury 2021). Water resource regulation includes many policies, strategies, and practices to ensure water system availability, quality, and resilience (Ferrier et al. 2020; Bogachev et al. 2023; Xiong et al. 2023; Rezvani et al. 2024). This chapter covers three interrelated aspects of water resources regulation: watershed management and climate change resilience, wetlands and flood regulation, and ecosystems' role in water quality. By analyzing these factors, we aim to demonstrate the importance of efficient regulation in protecting water resources for future generations.

Catchments, or drainage basins, are regions where water flows toward a river, lake, or ocean. Effective watershed management requires integrating land, water, and other natural resources within hydrological boundaries. Climate change causes changes in rainfall patterns, extreme weather events, and hydrological cycles, making water resource management difficult. Integrating land and water management can reduce floods and droughts (Choudhury et al. 2022). Conservation and restoration of native flora regulate water flow, reduce erosion, and replenish groundwater. Flexible water allocation systems and sustainable agriculture are essential to addressing climate change. Collaboration among stakeholders has improved water quality and ecological well-being by reducing nutrient pollution, restoring riparian buffers, and improving agricultural practices. The Australian government's National Water Initiative promotes watershed-level integrated water resource management for sustainable use, effectiveness, and environmental protection (Costello 2022; Dardonville et al. 2023; Sauterey et al. 2023; Hong et al. 2024). With their ecological value, wetlands face threats from urbanization, land development, industrial runoff pollution, and climate change. Future efforts should prioritize governance structures, stakeholder collaboration, and flexible management strategies to address these issues. Wetlands absorb excess water, protect against storm surges, and improve hydrological connectivity. Legal and regulatory measures are necessary for conservation and restoration efforts.

Several human activities and environmental stressors can degrade ecosystem water services. Deforestation and habitat destruction increase sedimentation and nutrient runoff into waterways (Tarolli et al. 2023; Aghaloo et al. 2024; Ranjan 2024; van Bommel et al. 2024; Zhang et al. 2024). Agriculture, industry, and urban pollution can introduce harmful substances into aquatic environments. Climate change and rising sea levels can disrupt ecological processes and water quality. Sustainable land management methods like agroforestry, cover cropping, and reduced tillage can improve water quality. Vegetation and buffer zones can create aquatic species habitats. Market-based approaches can motivate landowners to conserve and restore environmental services. However, challenges include conflicting land use interests and limited financial resources. Future efforts should focus on education, collaboration, and regulatory strategies to protect and enhance water resources (Fig. 3.3).

Biodiversity Conservation

Improved water quality and ecological function can maintain the planet's ecological balance and protect future generations. Ecosystems are robust because of Earth's complex network of species and their interactions, which provide essential services for human survival. This study examines the complex relationship between biodiversity, ecological resilience, and climate stability. We study the inherent relationships between biodiversity and ecological system resilience. We want to understand diverse habitats' ecological services and biodiversity risks. This has major implications for climate system resilience. Biodiversity is crucial to ecological system resilience (Shi et al. 2023; Berhanu et al. 2024; Kapruwan et al. 2024; Lv and Sarker 2024). According to the diversity-stability hypothesis, biodiversity increases ecological system stability and disturbance resilience. Diversity helps ecosystems adapt to environmental changes, ensuring their longevity and effectiveness. Functional diversity—the wide range of ecological roles and functions species perform in an ecosystem—is essential to biodiversity. Since many species perform similar ecological functions, ecosystems with diverse functional traits are more redundant. Without one species, others with similar functional characteristics can replace it, preserving ecosystem stability. The complex relationship between biodiversity and

Fig. 3.3 Water resources regulation. *Source* created by authors

ecological services shows how diverse habitats benefit humans. Diversity underpins many ecological services, including food, water, climate regulation, and disease control. Diverse ecological systems have a large genetic diversity reservoir, which offers resilience to environmental changes and ensures the continuity of vital ecological services needed for human survival. Many ecological systems provide vital services for human survival and economic well-being. Forests provide timber, fuelwood, and non-timber forest products for various societal needs. Agriculture relies on biodiversity for insect pollination and soil fertility from diverse microbial communities (Higuera Roa et al. 2023; Ross et al. 2024). Marine ecosystems support jobs and food security for many people worldwide in the fisheries sector. Biodiversity helps ecological systems regulate environmental equilibrium and reduce natural hazards. Protecting against floods and storms, wetlands reduce their impact. Biodiverse forests regulate atmospheric CO_2 and sequester carbon, mitigating climate change. Plant communities also stabilize soil, control erosion, and purify water, strengthening terrestrial ecosystems. Diverse ecosystems provide tangible benefits and intangible cultural and supporting services that improve human societies and well-being. Biodiversity-rich landscapes offer ecotourism, nature-based recreation, and spiritual renewal. Indigenous and local communities' cultural identity and traditional knowledge come from their interactions with ecological systems, demonstrating biodiversity's value beyond its practical use. Ecosystems also depend on supporting services like nutrient cycling, pollination, and pest regulation to sustain life and strengthen the environment. Deforestation, urbanization, and agricultural expansion threaten biodiversity urgently. Fragmentation disrupts ecological connectivity, isolating populations and reducing genetic diversity. This reduces ecosystem resilience. Depletion of tropical rainforests and coral reefs reduces biodiversity and weakens ecosystems' ability to store carbon and control climate, worsening climate change. Climate change alters species distribution, ecological systems, and extinction risks, making biodiversity conservation difficult. Increasing temperatures, shifting precipitation patterns, and severe weather events threaten species, especially those with limited habitat ranges and spread (Qin et al. 2021; Haldar et al. 2023; Aghaloo et al. 2024; Lv and Sarker 2024; Miao et al. 2024). Coral bleaching, caused by ocean warming, threatens marine biodiversity and the economies of coastal communities that depend on coral reef systems. The loss of key species and disruptions to food chain relationships make ecological systems more unstable, amplifying climate change's chain reaction effects on biodiversity. Overfishing, poaching, and illegal logging threaten biodiversity and ecological systems. Wildlife population extraction harms ecological equilibrium, reducing biodiversity and compromising ecological systems. Unsustainable fishing practises like bottom trawling and bycatch degrade marine habitats and reduce ecological resilience. Reduced keystone species, such as apex predators, can cause trophic cascades that degrade ecological systems and biodiversity. Invasive species and pollution make ecosystems more vulnerable. Invasive alien species outcompete native flora and fauna, disrupt ecological processes, and change community dynamics, resulting in biodiversity loss and habitat degradation. Pollution—including chemical contaminants, plastic waste, and nutrient runoff—degrades water quality, contaminates soils, and endangers wildlife and humans. Invasive species and pollution weaken ecological systems, making them more susceptible to climate change-related environmental stressors (Gogoi et al. 2024a, b) Biodiversity conservation is essential to ecosystem resilience and climate change mitigation. The complex interconnectedness of living organisms provides essential ecological services for human welfare and maintains ecosystems' resilience to environmental disruptions. However, biodiversity threats like habitat degradation and climate change threaten ecological systems and their ability to provide essential services. These challenges require local, national, and global collaboration to restore habitats, manage resources sustainably, and mitigate climate change. We can create a resilient and sustainable future for future generations by valuing biodiversity and protecting the environment (Fig. 3.4).

Ecosystem-Based Adaptation

Ecosystem-based adaptation (EbA) is an effective way to mitigate climate change. Using healthy ecosystems' resilience and services improves human and environmental welfare. EbA preserves, rehabilitates, and manages forests, wetlands, mangroves, and coral reefs. This method uses these environmental systems' inherent capacities to mitigate climate extremes like floods, droughts, and sea level rise.

Fig. 3.4 Biodiversity conservation. *Source* created by authors

Several fundamental principles guide EbA project implementation. Instead of fighting nature, EbA prioritizes cooperation. Healthy ecosystems are highly adaptable. Protecting and restoring these ecological systems increases their ability to provide essential human and environmental services.

Additionally, EbA promotes community-based methods. Local stakeholders plan, implement, and monitor EbA initiatives. This ensures that projects meet community needs and are culturally appropriate. The EbA perspective is comprehensive. According to the statement, ecological systems and human societies are interdependent (Kumar et al. 2022; Vilà-Cabrera et al. 2023; Yuan et al. 2024).

EbA improves ecological systems, which boosts biodiversity, water quality, and soil health. Ecological Balance Assessment (EbA) applications vary. Natural boundaries like mangrove forests protect coastal areas from storm surges and tsunamis. Restoration of degraded mangrove ecosystems can reduce coastal communities' flood and erosion risk. In heavy rain, wetlands absorb and store excess water. Restoration and protection of wetlands may reduce downstream flooding risks. Forest health affects water cycles. Sustainable forest management practices that maintain vegetation cover can preserve soil moisture and minimize drought damage. Integrating different plant species and optimizing soil conditions can boost agricultural resilience to climate change. Use cover cropping and crop rotation to retain soil moisture and reduce erosion. The examples above are just examples, and EbA implementations

Fig. 3.5 Ecosystem-based adaptation. *Source* Created by authors

are constantly expanding. Researchers are exploring new ways to use natural ecosystems to boost climate resilience. EbA works better when integrated into a climate resilience planning framework. Identifying a community's climate hazards requires a thorough vulnerability assessment. The assessment should evaluate nearby ecological systems. EbA strategies should improve and expand climate change adaptation plans. For instance, national and regional development policies incorporate EbA projects. EbA initiatives require strategic thinking to restore and manage ecosystems. Climate resilience planning must include the EbA project's long-term benefits and factors. An effective EbA intervention monitoring and evaluation program is essential. This helps make necessary changes and improvements. Communities can use natural resources to build a sustainable and resilient future by incorporating EbA into climate resilience planning. In this scenario, resilient ecological systems protect humans and thrive in a changing climate (Fig. 3.5).

Challenges and Opportunities

Eco-based approaches face many obstacles that hinder their widespread adoption and efficacy. Policy and governance are the main obstacles. Ecosystem-based approaches require comprehensive, cross-sectoral policies prioritizing environmental well-being and adaptability over economic benefits. However, current governance systems often prioritize specific sectors' interests, fragmenting decision-making and conflicting priorities. Transforming governance frameworks to promote collaboration, inclusivity, and sustainability is necessary to overcome this challenge. Lack of understanding and recognition of ecological services also hinders progress. Many decision-makers and stakeholders do not realize ecosystems' vital role in reducing human well-being and climate risk. Lack of consciousness often undervalues ecological services, causing their depletion. This challenge requires extensive educational and outreach efforts to raise awareness of ecosystems' value and benefits. By understanding ecological services, stakeholders can make informed decisions about conservation and restoration. Additionally, financial constraints hinder ecosystem-based methods. Although environmental restoration and conservation projects have long-term benefits, the initial costs may deter investment. However, payments for ecological services and green bonds can encourage investments in ecological system resilience. Public–private partnerships and blended finance can help ecosystem-based adaptation initiatives by mobilizing public and private funding. Technical capacity and knowledge gaps hinder ecosystem-based approaches. Successful environmental management requires knowledge and skills from ecology, hydrology, climatology, and social sciences. However, many regions lack technical and scientific knowledge, which hinders their ability to develop and implement ecological adaptation measures. Through training, knowledge exchange, and peer-to-peer learning networks, capacity-building initiatives empower local communities, policymakers, and practitioners (Tarolli et al. 2023; Berhanu et al. 2024; Zhang et al. 2024). These initiatives give them the skills to integrate ecosystem-based approaches into climate resilience strategies. Additionally, socio-cultural factors can hinder ecosystem-based methodologies. Traditional resource management, land ownership, and cultural beliefs can conflict with modern conservation principles, hindering cooperative efforts to protect and restore ecological systems. Respecting indigenous knowledge systems and local traditions is crucial when developing adaptation strategies that fit the context and local values and priorities. Stakeholder participation in participatory decision-making and traditional ecological knowledge can improve ecosystem-based approaches' social acceptability and efficacy. This approach fosters natural resource ownership and stewardship. Many opportunities exist to increase ecological contributions to climate resilience despite these obstacles. Nature-based solutions can mitigate climate change. Nature-based solutions use natural processes and diverse ecological systems to mitigate and adapt to climate change while improving resilience economically and environmentally. Restoring wetlands, forests, and mangroves can reduce flood risks, sequester carbon, increase biodiversity, and improve water quality and recreation. Additionally, ecosystem-based climate adaptation and disaster risk reduction strategies may increase resilience at multiple levels (Gogoi et al. 2024b). Policymakers can protect and restore natural systems to mitigate climate-related hazards by considering ecological factors in planning and decision-making. Infrastructure like urban green spaces and natural coastal defences can make cities and communities more resilient to extreme weather. These initiatives have social, economic, and environmental benefits. Technological advances can also improve ecosystem-based climate resilience strategies. Remote sensing, GIS, and machine learning algorithms help monitor and assess ecosystem health. These tools enable data-driven decision-making and targeted interventions. Digital platforms and mobile apps can also increase stakeholder engagement and enable participatory monitoring, empowering local communities to manage and conserve ecological systems. Multi-stakeholder partnerships and collaboration may also help expand and improve ecosystem-based approaches. Governments, civil society, academic institutions, and private companies can collaborate to restore and conserve ecological systems by combining their knowledge, resources, and networks. Landscape-level conservation programs and transboundary cooperation agreements may improve ecological connectivity and resilience across many ecological systems and jurisdictions. Ecosystem-based approaches require coordinated local, national, and global efforts to maximize their potential. Addressing policy fragmentation, financial constraints, and knowledge gaps while seizing opportunities for nature-based solutions, technological innovation, and multi-stakeholder collaboration can help communities adapt to climate change and protect their vital ecosystems. Ecologically oriented methods are necessary to build a sustainable and resilient future for future generations (Table 3.1).

Conclusion

In summary, ecosystems' significant role in enhancing climate resilience must be overemphasized and considered. In this chapter, we have explained the diverse mechanisms through which ecosystems function as inherent safeguards against the consequences of climate change. Ecosystems provide a wide range of services that contribute to the resilience of both natural and human systems, encompassing carbon sequestration in forests and flood regulation in wetlands. It is crucial to acknowledge that it is essential to prioritize the environment's resilience in climate adaptation strategies. Through protecting and rehabilitating ecological systems, we enhance our capacity to endure the detrimental impacts of climate change and establish the groundwork for sustainable development.

Moreover, the incorporation of ecosystem-based approaches into policy frameworks can result in various additional advantages, such as the preservation of biodiversity, the enhancement of livelihoods, and the improvement of social well-being. When considering the future, it becomes apparent that a significant requirement exists for ongoing research and the implementation of inventive policy interventions. In light of the ongoing challenges presented by climate change, it is imperative to investigate novel approaches aimed at bolstering ecological resilience and cultivating adaptive capacity. The effective development of strategies to harness the full potential of ecosystems in building climate resilience will heavily rely on collaborative efforts among scientists, policymakers, and local communities. Understanding and utilizing the inherent connections between ecosystems and climate resilience is crucial for effectively addressing the intricate challenges of the twenty-first century.

Table 3.1 Challenges and opportunities for ecosystems' contribution to climate resilience

Challenges	Opportunities	Advantages	Disadvantages	Mitigation strategies
Habitat fragmentation	Ecosystem restoration	• Enhances habitat connectivity	• Disruption of natural migration patterns	Establish wildlife corridors and restore degraded areas
Invasive species	Nature-based solutions	• Enhances biodiversity	• Out competition of native species	Monitoring and control programs, restoration efforts
Pollution	Green infrastructure	• Improves air and water quality	• Harmful effects on ecosystem health	Regulations, pollution control technologies
Overexploitation	Indigenous knowledge	• Supports sustainable resource use	• Depletion of natural resources	Implement sustainable harvesting practices
Land Use Change	Protected areas	• Preserves critical habitats	• Loss of natural landscapes	Zoning regulations, land-use planning
Deforestation	Sustainable land management	• Enhances carbon sequestration	• Loss of biodiversity, soil erosion	Reforestation, agroforestry practices
Desertification	Climate-smart agriculture	• Improves soil fertility	• Decreased land productivity	Soil conservation practices, drought-resistant crops
Ocean acidification	Marine protected areas	• Preserves marine biodiversity	• Coral bleaching, fishery decline	Reduce CO_2 emissions, coral reef restoration
Coral bleaching	Integrated water management	• Improves coastal resilience	• Loss of coral reefs	Reduce pollution and regulate fishing practices
Glacier retreat	Community-based conservation	• Fosters local stewardship	• Reduced freshwater supply	Glacier monitoring, water conservation initiatives
Sea level rise	Ecosystem-based adaptation	• Enhances coastal protection	• Habitat loss, inundation	Coastal zone planning, mangrove restoration
Extreme weather events	Carbon offsetting	• Supports carbon sequestration	• Disruption of ecosystem functions	Disaster preparedness, ecosystem restoration

(continued)

Table 3.1 (continued)

Challenges	Opportunities	Advantages	Disadvantages	Mitigation strategies
Disease Outbreaks	Climate Finance	• Mobilizes funds for conservation	• Decline in ecosystem health	Disease monitoring, habitat restoration
Loss of biodiversity	Sustainable fisheries management	• Maintains ecological balance	• Reduced ecosystem resilience	Marine protected areas, sustainable fishing
Permafrost thaw	Ecotourism	• Generates income for conservation efforts	• Release of greenhouse gases	Sustainable tourism practices, permafrost monitoring
Water scarcity	Green roofs and walls	• Reduces urban heat island effect	• Decreased water availability	Rainwater harvesting, water recycling
Soil degradation	Carbon farming	• Improves soil health	• Loss of soil fertility	Soil conservation practices, organic farming
Coastal erosion	Renewable energy	• Reduces dependence on fossil fuels	• Habitat loss, ecosystem disruption	Offshore wind farms, coastal restoration
Loss of wetlands	Urban greening	• Enhances urban biodiversity	• Loss of habitat for migratory birds	Wetland restoration, green infrastructure
Acid rain	Agroecology	• Promotes sustainable agriculture	• Soil and water acidification	Buffer zones, soil amendments
Land degradation	Soil remediation	• Restores soil fertility	• Reduced ecosystem productivity	Phytoremediation, erosion control measures
Ocean pollution	Circular economy	• Reduces waste generation	• Harmful effects on marine life	Waste management policies, marine cleanup efforts
Soil erosion	Regenerative agriculture	• Improves soil structure	• Loss of topsoil	Conservation tillage, cover cropping
Energy development	Green technology	• Promotes sustainable energy production	• Habitat destruction, pollution	Renewable energy projects, habitat restoration
Agricultural intensification	Sustainable forestry	• Maintains forest ecosystem services	• Deforestation, loss of biodiversity	Agroforestry, sustainable harvesting practices

(continued)

Table 3.1 (continued)

Challenges	Opportunities	Advantages	Disadvantages	Mitigation strategies
Urbanisation	Ecosystem services valuation	• Raises awareness about ecosystem benefits	• Habitat fragmentation, loss of green spaces	Innovative growth policies, green infrastructure
Loss of carbon sinks	Conservation finance	• Provides funding for conservation projects	• Increased carbon emissions	REDD+ initiatives, afforestation programs
Soil salinization	Drought-Resistant Crops	• Improves crop resilience	• Soil degradation, reduced productivity	Soil management practices, saline-tolerant crops
Coastal development	Blue carbon projects	• Enhances carbon sequestration in coastal ecosystems	• Habitat destruction, loss of biodiversity	Coastal zone management, habitat restoration
Overfishing	Sustainable aquaculture	• Provides alternative food sources	• Depletion of fish stocks, habitat degradation	Aquaculture certification, marine protected areas
Thermal pollution	Riparian buffers	• Improves water quality	• Harmful effects on aquatic life	Vegetative buffer strips, water treatment systems
Habitat degradation	Wildlife corridors	• Enhances habitat connectivity	• Fragmentation of ecosystems	Establish wildlife corridors, habitat restoration
Water pollution	Stormwater management	• Reduces runoff pollution	• Contamination of water sources	Green infrastructure, pollution control measures
Noise pollution	Biodiversity conservation	• Reduces disturbance to wildlife	• Disruption of animal behaviour	Noise barriers, habitat preservation
Soil contamination	Phytoremediation	• Removes contaminants from soil	• Soil and water pollution	Plant-based remediation techniques, soil monitoring
Loss of cultural ecosystem services	Environmental education	• Promotes cultural conservation	• Decline in cultural connections to nature	Community engagement, cultural heritage preservation

(continued)

Table 3.1 (continued)

Challenges	Opportunities	Advantages	Disadvantages	Mitigation strategies
Inequitable access to resources	Environmental justice	• Promotes equity in resource management	• Vulnerability to climate impacts	Advocacy for marginalised communities, policy reform
Governance challenges	Participatory decision-making	• Enhances transparency and accountability	• Ineffective resource management	Stakeholder engagement, capacity building
Lack of awareness	Science communication	• Increases public understanding of climate issues	• Limited support for conservation efforts	Outreach programs, education initiatives
Economic pressures	Sustainable consumption	• Promotes responsible resource use	• Overexploitation of natural resources	Green consumerism, sustainable development goals
Climate change denial	Climate literacy	• Fosters informed decision-making	• Hinders climate action efforts	Science-based education, communication strategies

We can establish a resilient and sustainable future generation by prioritizing and allocating resources towards our natural capital.

References

Aghaloo K, Sharifi A, Habibzadeh N et al (2024) How nature-based solutions can enhance urban resilience to flooding and climate change and provide other co-benefits: a systematic review and taxonomy. Urban for Urban Green 95:128320. https://doi.org/10.1016/j.ufug.2024.128320

Ahmad T, Kumar N, Kumar A et al (2024) Unveiling the potential of membrane in climate change mitigation and environmental resilience in ecosystem. Environ Res 245:117960. https://doi.org/10.1016/j.envres.2023.117960

Almulhim AI, Cobbinah PB (2024) Framing resilience in Saudi Arabian cities: on climate change and urban policy. Sustain Cities Soc 101:105172. https://doi.org/10.1016/j.scs.2024.105172

Azmi MA, Mokhtar K, Osnin NA et al (2023) Enhancing coastal ecosystem resilience: investigating the interplay between safety criteria and ferry employee's perceptions to address climate change impacts. Environ Res 238:117074. https://doi.org/10.1016/j.envres.2023.117074

Behboudian M, Anamaghi S, Mahjouri N, Kerachian R (2023) Enhancing the resilience of ecosystem services under extreme events in socio-hydrological systems: a spatio-temporal analysis. J Clean Prod 397:136437. https://doi.org/10.1016/j.jclepro.2023.136437

Berhanu AA, Ayele ZB, Dagnew DC (2024) Impact of climate-smart agricultural practices on smallholder farmers' resilience in Ethiopia. J Agric Food Res 16:101147. https://doi.org/10.1016/j.jafr.2024.101147

Bogachev MI, Gafurov AM, Iskandirov PY et al (2023) Reversal in the drought stress response of the Scots pine forest ecosystem: Local soil water regime as a key to improving climate change resilience. Heliyon 9:e21574. https://doi.org/10.1016/j.heliyon.2023.e21574

Choudhury M, Pervez A, Sharma A, Mehta J (2022) Human-induced stresses on the rivers beyond their assimilation and regeneration capacity. In: Ecological significance of river ecosystems. Elsevier, pp 281–298. https://doi.org/10.1016/B978-0-323-85045-2.00019-4

Costello MJ (2022) Biodiversity conservation through protected areas supports healthy ecosystems and resilience to climate change and other disturbances. In: DellaSala DA, Goldstein MIBT-ITE of C (eds). Elsevier, Oxford, pp 423–429

Dardonville M, Catarino R, Therond O (2023) Sustainability and resilience against climate change provided by a territorial crop-livestock system. J Clean Prod 432:139646. https://doi.org/10.1016/j.jclepro.2023.139646

Ferrier S, Harwood TD, Ware C, Hoskins AJ (2020) A globally applicable indicator of the capacity of terrestrial ecosystems to retain biological diversity under climate change: the bioclimatic ecosystem resilience index. Ecol Indic 117:106554. https://doi.org/10.1016/j.ecolind.2020.106554

Filippini M, Segadelli S, Dinelli E et al (2024) Hydrogeological assessment of a major spring discharging from a calcarenitic aquifer with implications on resilience to climate change. Sci Total Environ 913:169770. https://doi.org/10.1016/j.scitotenv.2023.169770

Gogoi N, Choudhury M, Ul Hasan MS, Changmai B, Baruah D, Samanta P (2024a) Soil organic carbon pool in diverse land utilization patterns in North-East India: an implication for carbon sequestration. Environ Dev Sustain, 1–27. https://doi.org/10.1007/s10668-024-05037-y

Gogoi N, Sarma A, Choudhury M, Samanta P, Faizan M, Sadhak S (2024b) Soil heavy metal pollution and ecological risk assessment in disturbed and undisturbed soil of Morigaon, Assam. Discover Environ 2(1):13. https://doi.org/10.1007/s44274-024-00039-z

Gong X, Wong W-K, Peng Y et al (2023) Exploring an interdisciplinary approach to sustainable economic development in resource-rich regions: an investigation of resource productivity, technological innovation, and ecosystem resilience. Resour Policy 87:104294. https://doi.org/10.1016/j.resourpol.2023.104294

Haldar S, Choudhury M, Choudhury S, Samanta P (2023) Trend analysis of long-term meteorological data of a growing metropolitan city in the era of global climate change. Total Environ Res Themes 7:100056. https://doi.org/10.1016/j.totert.2023.100056

Higuera Roa O, Cotti D, Aste N et al (2023) Deriving targeted intervention packages of nature-based solutions for climate change adaptation and disaster risk reduction: a geospatial multi-criteria approach for building resilience in the Puna region, Peru. Nature-Based Solut 4:100090. https://doi.org/10.1016/j.nbsj.2023.100090

Hong G, Liu S, Liu W, Wu X (2024) Nonlinear trade-off relationship and critical threshold between ecosystem services and climate resilience for sustainable urban development. Sustain Cities Soc 103:105253. https://doi.org/10.1016/j.scs.2024.105253

Kapruwan R, Saksham AK, Bhadoriya VS et al (2024) Household livelihood resilience of pastoralists and smallholders to climate change in Western Himalaya, India. Heliyon 10:e24133. https://doi.org/10.1016/j.heliyon.2024.e24133

Kumar R, Pande VC, Bhardwaj AK et al (2022) Long-term impacts of afforestation on biomass production, carbon stock, and climate resilience in a degraded semi-arid ravine ecosystem of India. Ecol Eng 177:106559. https://doi.org/10.1016/j.ecoleng.2022.106559

Kumar D, Choudhury M, Rathore A (2020) Valuation of ecosystem services & benefits of Son Beel wetland in Assam, India: a case study of natural solutions to climate change & water: the Worldwide Online Climate Change Conference. Hamburg University, Germany

Kumar D, Choudhury M (2021) Recognizing economic values of wetland ecosystem services: a study of emerging role of monetary evaluation of Chandubi ecosystem and biodiversity. In:

Wetlands conservation: current challenges and future strategies, pp. 87–110. https://doi.org/10.1002/9781119692621.ch5

Lacambra S CL, Spencer T, Munera C et al (2024) Coastal ecosystems contribution to climate adaptation and disasters risk management in the tropical Americas. Nat-Based Solut 5:100112. https://doi.org/10.1016/j.nbsj.2024.100112

Li X, Zhang M, Liu W, Wang Z (2024) The impact of entrepreneurial co-founders' support on founders' entrepreneurial resilience in the high-tech industry: the roles of perceived digital technology and perceived failure-tolerant climate. Technol Forecast Soc Change 202:123320. https://doi.org/10.1016/j.techfore.2024.123320

Liu Y, Li Q, Li W et al (2023) Knowledge map and hotspot analysis in climate resilience infrastructure (CRI) from 1997 to 2022 through scientometric analysis. Environ Res 228:115874. https://doi.org/10.1016/j.envres.2023.115874

Lv Y, Sarker MNI (2024) Integrative approaches to urban resilience: evaluating the efficacy of resilience strategies in mitigating climate change vulnerabilities. Heliyon 10:e28191. https://doi.org/10.1016/j.heliyon.2024.e28191

Maebe L, Dufrêne M, Claessens H et al (2023) The navigate framework: how the ecosystem services and resilience concepts can help us navigate in the current crises. Ecosyst Serv 64:101570. https://doi.org/10.1016/j.ecoser.2023.101570

Mallick M, Singh PK, Pandey R (2024) Harvesting resilience: tribal home-gardens as socio-ecological solutions for climate change adaptation and sustainable development in a protected area. J Clean Prod 445:141174. https://doi.org/10.1016/j.jclepro.2024.141174

Mehmood K, Anees SA, Rehman A et al (2024) Exploring spatiotemporal dynamics of NDVI and climate-driven responses in ecosystems: insights for sustainable management and climate resilience. Ecol Inform 80:102532. https://doi.org/10.1016/j.ecoinf.2024.102532

Meng Z, He M, Li X et al (2024) Spatio-temporal analysis and driving forces of urban ecosystem resilience based on land use: a case study in the Great Bay Area. Ecol Indic 159:111769. https://doi.org/10.1016/j.ecolind.2024.111769

Miao X, Li Z, Wang M et al (2024) Measurement of cultivated land ecosystem resilience in black soil region of Northeast China under the background of cultivated land protection policy in China: case study of Qiqihar City. J Clean Prod 434:140141. https://doi.org/10.1016/j.jclepro.2023.140141

Parker LM, Scanes E, O'Connor WA et al (2024) Resilience against the impacts of climate change in an ecologically and economically significant native oyster. Mar Pollut Bull 198:115788. https://doi.org/10.1016/j.marpolbul.2023.115788

Qiao D, Yuan W, Li H (2024) Regulation and resilience: panarchy analysis in forest socio-ecosystem of Northeast National Forest Region, China. J Environ Manage 353:120295. https://doi.org/10.1016/j.jenvman.2024.120295

Qin J, Hao X, Hua D, Hao H (2021) Assessment of ecosystem resilience in Central Asia. J Arid Environ 195:104625. https://doi.org/10.1016/j.jaridenv.2021.104625

Ranjan R (2024) Creating synergies between payments for ecosystem services, green bonds, and catastrophe insurance markets for enhanced environmental resilience. Land Use policy 136:106970. https://doi.org/10.1016/j.landusepol.2023.106970

Rezvani SMHS, Silva MJF, de Almeida NM (2024) NETOBRA: boosting urban resilience through a digital platform for the construction ecosystem. Procedia Struct Integr 55:72–79. https://doi.org/10.1016/j.prostr.2024.02.010

Rhymes JM, Evans D, Laudone G et al (2024) Biochar improves fertility in waste derived manufactured soils, but not resilience to climate change. Sci Total Environ 923:171387. https://doi.org/10.1016/j.scitotenv.2024.171387

Ross AG, Connolly K, Vögele S, Kuckshinrichs W (2024) A macro-level analysis of the socio-economic impacts of climate change driven water scarcity: incorporating behavioural and resilience aspects. Water Res X 100223. https://doi.org/10.1016/j.wroa.2024.100223

Sauterey B, Le Gland G, Cermeño P et al (2023) Phytoplankton adaptive resilience to climate change collapses in case of extreme events—a modeling study. Ecol Modell 483:110437. https://doi.org/10.1016/j.ecolmodel.2023.110437

Sealey KS (2024) Assessing coastal vulnerability and climate resilience for Caribbean small island states using ecological principles. Int J Disaster Risk Reduct 105:104410. https://doi.org/10.1016/j.ijdrr.2024.104410

Shi X, Zhang Y, Ding H et al (2023) Drought risk assessment considering ecosystem resilience: a case study in the Huang-Huai-Hai Plain, China. Ecol Indic 156:111102. https://doi.org/10.1016/j.ecolind.2023.111102

Singh S, Singh P, Sharma A, Choudhury M (eds) (2022) Agriculture waste management and bioresource: the circular economy perspective. Wiley. https://onlinelibrary.wiley.com/doi/book/10.1002/9781119808428

Singh S, Goyal MK (2023) Enhancing climate resilience in businesses: the role of artificial intelligence. J Clean Prod 418:138228. https://doi.org/10.1016/j.jclepro.2023.138228

Sharma A, Choudhury M (2021) Changing climate and depleting water resources in the mountains with a case study from the Himalayas. In: Water conservation in the era of global climate change. Elsevier, pp 411–436. https://doi.org/10.1016/B978-0-12-820200-5.00013-0

Tarolli P, Zuecco G, Picco L (2023) Watershed resilience to climate change: from hillslopes to rivers and coasts. Evol Earth 1:100010. https://doi.org/10.1016/j.eve.2023.100010

Trégarot E, D'Olivo JP, Botelho AZ et al (2024) Effects of climate change on marine coastal ecosystems—a review to guide research and management. Biol Conserv 289:110394. https://doi.org/10.1016/j.biocon.2023.110394

van Bommel N, Höffken JI, Chatterjee I (2024) Building climate resilience through energy access? An empirical study on grid connectivity in the Indian Sundarbans. Energy Res Soc Sci 112:103504. https://doi.org/10.1016/j.erss.2024.103504

Vilà-Cabrera A, Astigarraga J, Jump AS et al (2023) Anthropogenic land-use legacies underpin climate change-related risks to forest ecosystems. Trends Plant Sci 28:1132–1143. https://doi.org/10.1016/j.tplants.2023.04.014

Wood SLR, Dupras J (2021) Increasing functional diversity of the urban canopy for climate resilience: potential tradeoffs with ecosystem services? Urban for Urban Green 58:126972. https://doi.org/10.1016/j.ufug.2020.126972

Xie W, Sun C, Lin Z (2023) Spatial-temporal evolution of urban form resilience to climate disturbance in adaptive cycle: a case study of Changchun city. Urban Clim 49:101461. https://doi.org/10.1016/j.uclim.2023.101461

Xiong M, Zhao R, Li X et al (2023) Climate resilience assessment of sustainability at national level: a case study of sub-Saharan Africa. J Clean Prod 430:139717. https://doi.org/10.1016/j.jclepro.2023.139717

Yuan Y, Zheng Y, Huang X, Zhai J (2024) Climate resilience of urban water systems: a case study of sponge cities in China. J Clean Prod 451:141781. https://doi.org/10.1016/j.jclepro.2024.141781

Zhang Z, Wang Y, Liang C et al (2024) Seasonal resilience of temperate estuarine fish in response to climate change. Ecol Indic 158:111518. https://doi.org/10.1016/j.ecolind.2023.111518

Chapter 4
Ecosystem Resilience for Climate Resilience, Strategies and Innovations: A Case Study from Developing Countries

Donizete Beck, Anindita Pal, Purvi Zaveri, Anran Zheng, Atul K. Tiwari, and Sushobhan Majumdar

Abstract This chapter examines the strategies and innovations necessary for enhancing ecosystem resilience to climate change, focusing on developing countries. It highlights the importance of biodiversity and structural diversity in increasing resilience, suggesting that management practices promoting heterogeneity can mitigate climate warming impacts. Ecosystem restoration is highlighted as a critical strategy involving identifying and removing stressors such as pollution and land degradation. Techniques like plant establishment, topsoil replacement and reestablishment of microbial flora are essential for restoring ecosystem health and ensuring long-term recovery. Capacity building through educational programs is crucial for promoting sustainable practices and fostering climate-resilient ecosystems. Higher education institutions play a significant role in embedding sustainability principles and encouraging interdisciplinary collaboration despite challenges like limited resources. Technological solutions, particularly AI, are emphasized for their potential to improve climate predictions and support proactive adaptation strategies. These technologies can enhance decision-making strategies. While the theoretical underpinnings of ecosystem resilience are extensively documented, there still needs to be a more practical application of effective strategies within resource-constrained environments, necessitating the promotion of customized methodologies to tackle

D. Beck
Insper Institute of Education and Research, Sao Paulo, Brazil

A. Pal · A. K. Tiwari (✉)
Department of Geography, The Maharaja Sayajirao University of Baroda, Gujarat, India
e-mail: akt.geobhu@gmail.com

P. Zaveri
Biocare Research (I) Pvt. Ltd., Sustainability and Innovation Co Chair, YPFP, Tokyo, Japan

A. Zheng
AECOM Technical Services Inc., Dallas, TX, USA

S. Majumdar
Department of Geography, Pandit Raghunath Murmu Smriti Mahavidyalaya, Bankura, India

M. Choudhury et al. (eds.), *Sustainable Synergy: Harnessing Ecosystems for Climate Resilience*, Climate Change Management,
https://doi.org/10.1007/978-3-031-77957-2_4

the distinct obstacles encountered by developing nations. Obstacles to the implementation process, such as inadequate education, insufficient infrastructure, and the lack of localized data, are acknowledged as significant challenges. Addressing these challenges necessitates a dedication to capacity building and promoting accessible technological innovations. This paper advocates for a multifaceted approach that integrates ecosystem restoration, capacity building, technological solutions, and community-based strategies to enhance ecosystem resilience and climate change mitigation efforts in developing countries.

Keywords Ecosystem resilience · Climate resilience · Climate change · Capacity building · Resource management

Introduction

In the face of rapid environmental changes and increasing anthropogenic pressures, understanding and enhancing ecosystem resilience has become a significant concern for sustainable development (Folke et al. 2010; Zickgraf et al. 2022). Ecosystem resilience, the capacity of ecosystems to withstand and recover from disturbances, is crucial for maintaining biodiversity and ecosystem services that are vital for human survival and well-being (Kumar and Choudhury 2021). Developing countries, in particular, face significant challenges in maintaining ecosystem resilience due to limited resources and socio-economic constraints (Leão et al. 2017). Addressing these challenges is essential for fostering sustainable development worldwide. The construct of ecosystem resilience has attracted considerable attention from the scientific community and policymakers. As complex and dynamic entities, ecosystems are characterized by intricate interactions among living organisms and their physical environment (Kumar 2010). These interactions enable ecosystems to absorb disturbances, reorganize, and continue functioning without collapsing into an altered state (Park et al. 2023). It has been shown that resilient ecosystems, by maintaining these functions and structures, can better cope with environmental changes, providing stable ecosystem services such as clean air and water, climate regulation, and soil fertility (Gogoi et al. 2024a, b; Gogoi et al. 2024a, b; Almenar et al. 2021; Costanza et al. 2017; Khan 2021; Potschin and Haines-Young 2016).

Thus, the purpose of our review is twofold: first, to explore the concept of ecosystem resilience, and second, to identify strategies and innovations for enhancing ecosystem resilience in developing countries. This review examines the existing literature and highlights successful case studies to provide a better understanding of ecosystem resilience and its implications for sustainable development in the Global South.

Conceptual Framework

Ecosystem Resilience

Ecosystems are communities of living organisms (e.g., animals, plants, and microbes) interacting with their physical environment (e.g., air, water, and mineral soil) as a system (Chapin et al. 2011). These interactions involve the flow of energy and the cycling of nutrients, creating a balanced and self-sustaining environment. Importantly, different ecosystems, known as panarchy, are intricately interconnected, even those that seem not to have any relationship, such as the terrestrial and aquatic ecosystems (Gunderson and Holling 2002; Scharlemann et al. 2021). Local communities, scholars, policymakers, and global governance play a crucial role in resolving the conflict between human activities and natural issues to foster sustainable development. Collective action is essential for preserving our ecosystems (Beck 2023; Barnaud et al. 2018).

In this context, ecosystems support a wide variety of life forms, contributing to the planet's biodiversity, which is crucial for the stability and resilience of our landscapes and life on Earth (Levin 1999; Manning et al. 2018; Karsai et al. 2020). Ecosystems provide essential services (e.g., clean air and water, pollination of crops, climate regulation, and soil fertility), which are vital for human survival and well-being as well as other species (Almenar et al. 2021; Bennett et al. 2009; Bolund and Hunhammar 1999; Costanza et al. 2017; Fisher et al. 2009; Koellner 2011; Khan 2021; Nelson et al. 2009; Potschin and Haines-Young 2016; Zickgraf et al. 2022). These services are known as ecosystem services, which also must be resilient. See the work of Biggs et al. (2012) for a rich explanation of the seven principles for resilient ecosystem services. Ecosystems also offer economically valuable resources, such as timber, fish, and medicinal plants (Daily and Ruckelshaus 2022).

Climate Resilience

Addressing climate change is crucial for a sustainable future, including the interrelation between the climate and ecosystem complexities (Cassin et al. 2021; Hayes et al. 2019). Climate resilience appears in this context. According to the Intergovernmental Panel on Climate Change (2022, p. 7), climate resilience is the capacity/ability of ecosystems "to cope with a hazardous event or trend or disturbance, responding or reorganising," and simultaneously maintaining their essential function, identity, structure, biodiversity, and the capacity for adaptation, learning, and transformation. Ecosystem and climate resilience are deeply interconnected concepts, reinforcing each other (Folke et al. 2010; Gunderson and Holling 2002).

Methodologies and Approaches

Literature Search Methodology

Resilience to natural hazards requires multidisciplinary approach involving economic, governance, engineering, and environmental science perspectives. Furthermore, a systematic review of empirical studies on social-ecological systems reveals that many studies fail to fully operationalize resilience, often needing clear system boundaries and effective integration of social and ecological dimensions, which are crucial for a holistic understanding of ecosystem resilience (Tuan 2024).

Data Synthesis

Synthesizing qualitative findings on climate resilience poses methodological challenges and opportunities. Integrating diverse methods, languages, and knowledge systems is crucial for projects like ARCPATH. Traditional tools may struggle to capture community perspectives. Selective participation and missing data in longitudinal studies complicate understanding resilience. AI and Big Data offer opportunities for dynamic climate risk management platforms (Paulauskaite-Taraseviciene et al. 2022).

Strategies for Enhancing Ecosystem Resilience

Natural Resource Management

Ecosystem resilience depends on factors like the spatial and temporal scale of the disturbance, the presence of diverse flora and fauna, their characteristics, community organizational structure and their interplay with the disturbance. With the overutilization of natural resources, the world has already crossed six out of nine planetary boundaries. With that advanced understanding and influence of relationships between the ecosystems functioning, conserving diversity and sustaining ecosystems' resilience can help manage natural resources.

Sustainable resource management is finding a critical balance between resource availability and requirement by the community, which climate change challenges. Any approach for Sustainable Land Management (SLM) or Water Management (SWM) needs to account for resource management's financial and social aspects to be successfully applicable in developing countries. With the idea of tapping into indigenous knowledge and technological advancements at the same time, a global database for Sustainable Land Management (SLM) was developed by the United

Nations Convention to Combat Desertification (UNCCD). This database contains a unique set of 2439 (as of 8th Aug 2024) practices/technologies/approaches being registered from 139 countries. Various technologies and approaches like crop rotation, fallowing, intercropping, tillage reduction, construction of shelter belts, gully healing, etc., have been identified as popular methods local communities use in developing countries (FAO 2016). With increased momentum towards urban living, conservation strategies for biodiversity are proposed through green corridors or wildlife corridors across city areas (Zellmer and Goto 2022). Similarly, steps for protecting biodiversity-rich regions by developing National parks such as Kainji National Park in Nigeria and Kaziranga National Park in India play a critical role in conserving ecosystem functions (Das 2017; Tang and Adesina 2022).

Climate-Smart Agriculture

Agroecological practices integrate ecological practices with agricultural production techniques. With limited land available for food production and an increasing population, agroecological practices like crop diversification and intercropping, which enhance soil fertility, pest control, and nutrient cycling (FAO 2016), give hope for developing sustainable climate-smart agriculture.

Ecosystem Restoration

Climate change has affected each type of ecosystem in more than one way. Restoring natural capital requires intense rehabilitation activities. As part of the Paris Climate Accord, 196 nations ratified the Convention on Biological Diversity, and 137 nations included restoration in their plans, which indicate the urgency and need for restoring the ecosystem at various levels (Strassburg et al. 2020).

The dynamics of the restoration of water bodies are different. Water bodies could be stagnant water bodies (Groundwater or lakes), or they can be dynamic (like rivers or oceans). Dynamic water bodies are found to dilute the pollutants but also allow them to percolate or be carried to different ecosystems (Zaveri et al. 2020). Reports suggest that the successful restoration of rivers has been reported by IUCN (Stephenson 2014) by restoring lateral and longitudinal connectivity and vegetation and reconstructing channels for the rivers of Ireland and the UK. However, for developing countries like Columbia, data-related limitations pose challenges to river restoration planning and execution.

Innovations and Technologies

Technological Solutions

Geographic Information Systems (GIS) can play a crucial role in identifying and mitigating the impacts of human activities on ecosystems by providing spatial analysis and visualization tools that integrate various environmental data. For instance, GIS can be used to monitor environmental pressures in urban areas by collecting and analyzing indicators such as air quality, traffic, and waste, as demonstrated in Tehran, where the enhanced Urban Carrying Capacity Load Number (EUCCLN) model was employed to assess the degree of human-induced pressures across different districts (Boahen et al. 2023). Figure 4.1 shows ecosystem resilience for climate resilience.

Fig. 4.1 Ecosystem resilience for climate resilience

Community-Based Approaches

Indigenous knowledge and local practices are crucial in enhancing ecosystem resilience amidst climate change by leveraging centuries-old wisdom and sustainable resource management techniques (Amponsah et al. 2024). Indigenous communities possess a deep understanding of their environments, which includes knowledge of weather patterns, ecological indicators, and sustainable practices such as agroforestry, rotational farming, and water conservation (Ryplova et al. 2023). These practices mitigate environmental degradation and bolster community resilience and well-being by integrating social, cultural, and ecological dimensions of sustainability. For instance, the Pashtun pastoral community in Pakistan relies on their culturally embedded knowledge and social structures, such as kinship bonds and reciprocity, to strengthen their resilience against climate impacts (Narura et al. 2024). Similarly, the Indigenous communities in Botswana utilize their ecological knowledge to understand climate change impacts and adapt their livelihoods accordingly. However, there is a need to harmonize this knowledge with scientific methods for more effective climate strategies (Easton-Gomez et al. 2022). Figure 4.2 shows community based approaches and ecosystem resilience.

Ecosystem Resilience for Climate Resilience, Case Study from Developing Countries

Case Study of India

The Vulnerability-Resilience Indicator Prototype (VRIP) assessment provides significant insights into the effects of climate change in India (Brenkert and Malone 2005). It identifies nine states with moderate resilience due to low sulfur emissions and unmanaged land challenges, which alleviate some environmental impacts. Conversely, six states are deemed more vulnerable than average due to geographical risks associated with sea storm surges, necessitating focused interventions in these areas. Regional assessments in states like Orissa and Maharashtra demonstrate the VRIP's effectiveness when paired with socio-economic data. The research highlights the necessity of incorporating socio-economic and environmental indicators within the VRIP framework to evaluate vulnerability and resilience comprehensively. This holistic approach is vital for understanding the broader context of climate impacts. Ultimately, the VRIP assessment advocates for differentiated strategies to tackle climate change effects in India, emphasizing that effective management and mitigation must be tailored to each state's specific environmental and socio-economic contexts.

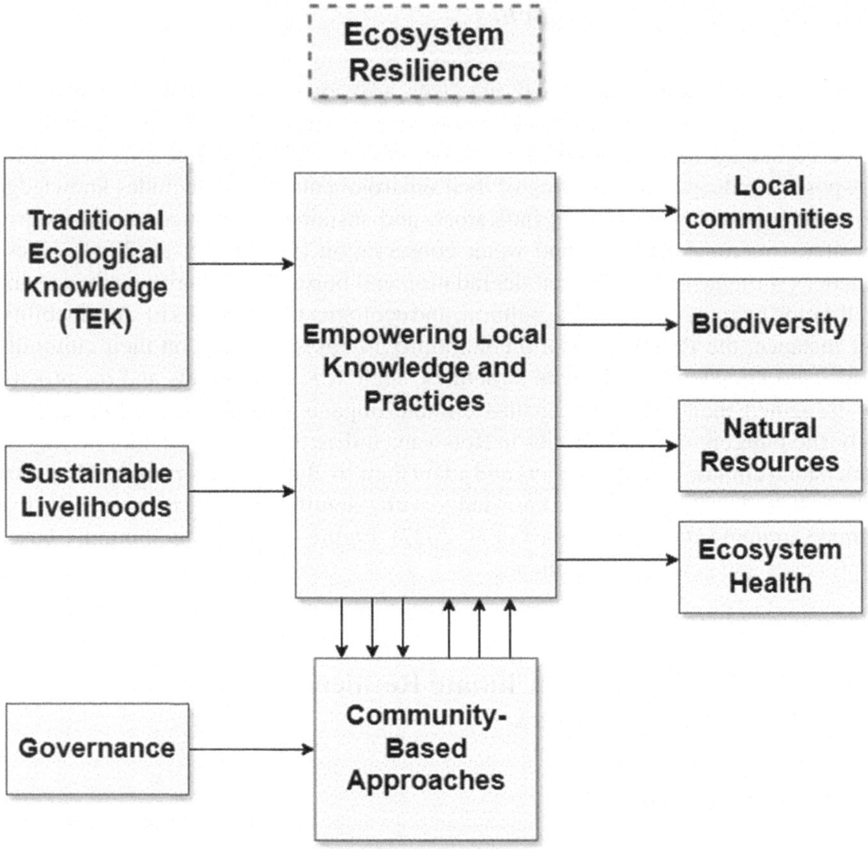

Fig. 4.2 Community based approaches and ecosystem resilience

Case Study of Sub Saharan Africa

The inadequate representation of urban ecosystem services in research emphasizes the need for a comprehensive ecological, economic and socio-cultural understanding (Mngumi 2020). It aims to identify research gaps in urban and peri-urban ecosystem services, particularly in Sub-Saharan Africa (SSA), to stimulate further investigation and application for climate resilience and sustainable development. Research on ecosystem services in SSA exhibits a notable rural bias, indicating the necessity for urban-focused studies to address pressing environmental challenges. The review reveals that urban and peri-urban areas in SSA remain underexplored despite advancements in global ecosystem services research, indicating a need for a paradigm shift. It explores how different ecosystem services, such as provisioning and regulating services, can contribute to climate change mitigation and adaptation efforts.

The paper underscores the importance of integrating ecosystem services into governance frameworks for peri-urban areas, which often face fragmented management and planning challenges. It advocates for a holistic approach that recognizes the interconnections between urban and rural areas to promote sustainable development and ecosystem service delivery. Significant barriers to realizing ecosystem service benefits in urban and peri-urban regions include vague interpretations of the concept, limited research focus, and considerable ecosystem degradation.

Case Study of Sub Saharan Brazil

The urgent need for ecosystem restoration in Brazil is due to critical land degradation impacting climate, biodiversity, and sustainable development (Bustamante et al. 2019). It highlights the role of well-managed ecosystems in climate hazard mitigation and integrating land use policy with restoration to meet international climate goals, particularly the Paris Agreement. Degraded landscapes in Brazil and restoration demand Brazil's native vegetation covers approximately 530 million hectares, with agricultural expansion as a primary cause of degradation. Leading to significant carbon emissions and loss of ecosystem services, the chapter identifies obstacles to restoring 12 million hectares of forest by 2030, such as logistical challenges, exotic species risks, and the need for a phased restoration approach. It stresses the importance of understanding local ecological contexts and stakeholder perspectives to devise effective restoration strategies that reconcile ecological and economic factors. The estimated cost for restoring 12 million hectares of forest is around $15.3 billion, with potential economic benefits including job creation and improved ecosystem services. Successful large-scale restoration initiatives hinge on effective governance and policy coordination, requiring stakeholder collaboration and the integration of environmental policies. The paper calls for streamlining bureaucratic processes to promote the development and dissemination of restoration practices. The potential adverse effects of afforestation, particularly involving exotic species on water resources and biodiversity, are analyzed, drawing lessons from global experiences.

Case Study of Sub Saharan Bhutan

The impacts of climate change on Bhutan's highland communities, particularly regarding caterpillar fungus and yak farming (Dorji et al. 2023). It highlights rising temperatures in mountainous regions that threaten agricultural productivity and food security. Study Site The research focuses on Laya and Lunana, highland communities dependent on yak herding and caterpillar fungus, facing harsh climatic conditions and limited resources. These challenges negatively affect their economic activities and livelihoods. Due to Climate Change, Economic activities differ markedly, with Laya exploring various income sources while Lunana primarily relies on caterpillar fungus. Climate change adversely affects caterpillar fungus income, prompting migration and shifts in traditional practices. Many respondents reported discernible temperature changes, negatively influencing glacier melt and precipitation. Communities observed alterations in rangeland species and heightened vulnerability to natural disasters, directly affecting their livelihoods. Botanical Species Composition and Emergence of Diseases Deteriorating caterpillar fungus quality and yield raise concerns, with new insect vectors emerging that threaten livestock and human health. Adaptation strategies include diversifying income through tourism and non-wood forest products, although challenges remain due to access and market limitations. The study clarifies the complex interplay between climate change, economic shifts, and traditional practices in highland communities. Finally, the study advocates for sustainable development strategies to support highland cultures and livelihoods amid ongoing environmental changes.

Challenges and Limitations

Barriers to Implementation

Socioeconomic constraints and policy gaps often hinder the implementation of strategies for enhancing ecosystem resilience for climate adaptation. Financial limitations become a critical challenge, as many communities, especially in developing regions, need more resources to invest in ecosystem restoration, conservation, and sustainable management practices. This is compounded by poverty and social inequality, which limit the capacity of vulnerable populations to adapt to climate impacts or engage in resilience initiatives. Cultural practices and traditional livelihoods may also resist changes, making it challenging to introduce new strategies (Santos et al. 2023). Additionally, many climate adaptation policies prioritize short-term economic benefits over long-term sustainability, neglecting the importance of ecosystem-based approaches (Tanner et al. 2017). Weak enforcement and inadequate stakeholder engagement, mainly the exclusion of local communities and private sector players, further impede the successful implementation of resilience strategies (Collier 2015). These barriers highlight the need for more coordinated, inclusive, and well-supported approaches to building ecosystem resilience.

Besides the socioeconomic and policy barriers, technological and knowledge limitations become another obstacle to implementing sustainable strategies. A significant challenge is the lack of access to advanced technologies that are essential for monitoring, modelling, and managing ecosystems effectively (Wang et al. 2023).

Adaptation and Resilience Building

Adaptive management approaches and enhancing adaptive capacity at local and national levels are critical for successfully implementing ecosystem resilience strategies. Adaptive management, a dynamic and iterative process, involves continuously monitoring ecosystems, assessing the effectiveness of strategies, and making adjustments based on new information and changing conditions (Birgé et al. 2016). This approach allows for flexibility in decision-making, enabling communities and governments to respond to unexpected environmental changes and evolving climate impacts (Schultz et al. 2015). Enhancing adaptive capacity involves building the skills, knowledge, and resources needed to implement these strategies effectively (Angeler et al. 2019).

Policy Implications and Recommendations

Policy Frameworks

Critical indicators of ecosystem resilience for climate adaptation policy-making are structural diversity, ecological fragility, strategic relevance, and tipping point likelihood. Structural diversity, both horizontal and vertical, enhances ecosystem resilience. Ecological fragility gauges an ecosystem's ability to absorb and recover from disturbances, identifying vulnerabilities (Tripathy et al. 2024). The strategic relevance of ecosystems highlights their role in supporting biodiversity and services. Understanding resilience about external drivers is crucial for ecosystems approaching tipping points. The Cusp Resilience Assessment (CUSPRA) model estimates tipping point likelihood, aiding ecosystem management. Developing and monitoring formal indicators for adaptation and resilience is essential (Singletary et al. 2022). Data quality gaps at the local level pose challenges for adaptation efforts. A unified approach and tailored local strategies are recommended for effective planning. By integrating these indicators—structural diversity, ecological fragility, strategic relevance, and tipping point likelihood—into climate adaptation policies, decision-makers can better support the resilience and sustainability of ecosystems in the face of global climate change (Atkinson and Alibašić 2023).

Capacity Building

Educational programs are essential for promoting sustainable practices by providing knowledge, skills, and ethical frameworks. Climate change education is crucial for increasing awareness and critical thinking to address climate change. Higher education institutions are instrumental in nurturing responsible global citizens by embedding sustainability principles across disciplines and promoting interdisciplinary collaboration despite challenges such as limited resources and institutional resistance. Furthermore, integrating sustainability in educational programs contributes to the Sustainable Development Goals (SDGs) by fostering conscious and active citizens better equipped to tackle environmental and social challenges (Abbaszadeh Tehrani et al. 2023). The involvement of stakeholders like managers, staff, and students is vital for the success of sustainability initiatives. Therefore, educational programs are fundamental in shaping future generations capable of contributing to climate-resilient ecosystems through informed decision-making and sustainable practices (Akinsete et al. 2022).

Supporting research and development of innovative solutions in ecosystem resilience for climate change mitigation requires a multifaceted approach that integrates various strategies and insights from different fields (Landa-Ochoa et al. 2024). A comprehensive framework can identify factors affecting innovation ecosystem resilience. Policymakers can use this framework to choose appropriate policies. Using climatological insights for sustainable natural resource management is essential. By integrating advanced climatic models with eco-innovative practices, we can predict and mitigate the impacts of climate variability on natural resources, thereby fostering environmental stability and resilience (Schneider et al. 2024). Nature-based solutions enhance urban resilience to climate change. More studies are required to assess these solutions' scalability and long-term effects.

Furthermore, the fusion of weather forecasting with adaptive resilience strategies can provide timely and accurate information, enabling better adaptation planning across various sectors such as infrastructure, agriculture, and communities (Tanguay et al. 2023). Lastly, interdisciplinary collaboration and investment in eco-innovation are essential for securing a resilient and sustainable future. By addressing technological, economic, and policy considerations, we can overcome the challenges and harness the opportunities presented by eco-innovative strategies (Ossewaarde et al. 2021). Collectively, these strategies provide a robust foundation for supporting research and development in ecosystem resilience, ultimately contributing to effective climate change mitigation.

Conclusion

Ecosystem resilience, the ability of ecosystems to absorb disturbances and reorganize while maintaining their core functions and structure, is essential for effective conservation and management. Key indicators such as biodiversity, functional diversity, connectivity, and adaptive capacity are critical in guiding management practices to enhance resilience, like protecting keystone species or introducing species with high adaptive potential. This approach safeguards biodiversity and preserves vital ecosystem services for human well-being. However, implementing these strategies in developing countries faces significant challenges due to economic constraints, social inequalities, and governance issues. Financial limitations often prioritize short-term economic gains over long-term ecological sustainability, while socio-economic disparities hinder community engagement, and weak governance impairs the enforcement of environmental regulations. Addressing these challenges requires increased financial investment, better community involvement, and more robust governance frameworks to ensure sustainable ecosystem management.

References

Abbaszadeh Tehrani N, Farhanj F, Janalipour M (2023) Monitoring the impacts of human activities on urban ecosystems based on the enhanced UCCLN (EUCCLN) model. ISPRS Int J Geo-Inf 12(4). https://doi.org/10.3390/ijgi12040170

Akinsete E, Stergiopoulou L, El Said N, Koundouri P (2022) Multi-actor working groups as fora for WEF nexus innovation and resilience. Environ Sci Proc 69. https://doi.org/10.3390/environsciproc2022015069

Almenar JB, Elliot T, Rugani B, Philippe B, Gutierrez TN, Sonnemann G, Geneletti D (2021) Nexus between nature-based solutions, ecosystem services and urban challenges. Land Use Policy 100:104898. https://doi.org/10.1016/j.landusepol.2020.104898

Amponsah M, Maryani E, Nandi N (2024) Systematic literature review on enhancing disaster resilience through spatial planning strategies. Int J Multidiscip Res 6(2). https://doi.org/10.36948/ijfmr.2024.v06i02.13156

Angeler DG, Fried-Petersen HB, Allen CR, Garmestani A, Twidwell D, Chuang WC, Donovan VM, Eason T, Roberts CP, Sundstrom SM, Wonkka CL (2019) Adaptive capacity in ecosystems. In: Advances in ecological research, vol 60, pp 1–24. Academic Press

Atkinson CL, Alibašić H (2023) Prospects for governance and climate change resilience in peatland management in Indonesia. Sustainability (Switzerland) 15(3). https://doi.org/10.3390/su15031839

Barnaud C, Corbera E, Muradian R, Salliou N, Sirami C, Vialatte A, Choisis P, Dendoncker N, Mathevet R, Moreau C, Reyes-García V, Boada M, Deconchat M, Cibien C, Garnier S, Maneja R, Antona M (2018) Ecosystem services, social interdependencies, and collective action: a conceptual framework. Ecol Soc 23(1). https://doi.org/10.5751/ES-09848-230115

Beck D (2023) Stakeholder theory for sustainable cities and society: a humanist and environmental approach for integrating people, institutions, and environmental ecosystems. Studia Ecologiae et Bioethicae 21(1):59–67. https://doi.org/10.21697/seb.2023.03

Bennett EM, Peterson GD, Gordon LJ (2009) Understanding relationships among multiple ecosystem services. Ecol Lett 12(12):1394–1404. https://doi.org/10.1111/j.1461-0248.2009.01387.x

Biggs R, Schlüter M, Biggs D, Bohensky EL, BurnSilver S, Cundill G, Dakos V, Daw TM, Evans LS, Kotschy K, Leitch AM (2012) Toward principles for enhancing the resilience of ecosystem services. Annu Rev Environ Resour 37(1):421–448. https://doi.org/10.1146/annurev-environ-051211-123836

Birgé HE, Allen CR, Garmestani AS, Pope KL (2016) Adaptive management for ecosystem services. J Environ Manage 183:343–352

Boahen S, Oviroh PO, Austin-Breneman J, Miyingo EW, Papalambros PY (2023) Understanding resilience of agricultural systems: a systematic literature review. Proc Des Soc 3(JULY):3701–3710. https://doi.org/10.1017/pds.2023.371

Bolund P, Hunhammar S (1999) Ecosystem services in urban areas. Ecol Econ 29(2):293–301. https://doi.org/10.1016/S0921-8009(99)00013-0

Brenkert AL, Malone EL (2005) Modeling vulnerability and resilience to climate change: a case study of India and Indian states. Clim Change 72(1–2):57–102. https://doi.org/10.1007/s10584-005-5930-3

Bustamante MMC, Silva JS, Scariot A, Sampaio AB, Mascia DL, Garcia E, Sano E, Fernandes GW, Durigan G, Roitman I, Figueiredo I, Rodrigues RR, Pillar VD, de Oliveira AO, Malhado AC, Alencar A, Vendramini A, Padovezi A, Carrascosa H, Freitas J, Siqueira JA, Shimbo J, Generoso LG, Tabarelli M, Biderman R, de Paiva Salomão R, Valle R, Junior B, Nobre C (2019) Ecological restoration as a strategy for mitigating and adapting to climate change: lessons and challenges from Brazil. Mitig Adapt Strat Glob Change 24(7):1249–1270. https://doi.org/10.1007/s11027-018-9837-5

Cassin J, Matthews JH (2021) Nature-based solutions, water security and climate change: Issues and opportunities. In: Cassin J, Matthews JH, Gunn EL (eds) Nature-based solutions and water security: an action agenda for the 21st century. Elsevier, Amsterdam, pp 63–79. https://doi.org/10.1016/B978-0-12-819871-1.00017-8

Chapin FS, Matson PA, Mooney HA, Vitousek PM (2011) Principles of terrestrial ecosystem ecology. Springer, New York. https://doi.org/10.1007/978-1-4419-9504-9

Collier MJ (2015) Novel ecosystems and social-ecological resilience. Landscape Ecol 30:1363–1369

Costanza R, De Groot R, Braat L, Kubiszewski I, Fioramonti L, Sutton P, Farber S, Grasso M (2017) Twenty years of ecosystem services: how far have we come and how far do we still need to go? Ecosyst Serv 28:1–16. https://doi.org/10.1016/j.ecoser.2017.09.008

Daily GC, Ruckelshaus M (2022) 25 years of valuing ecosystems in decision-making. Nature 606(7914):465–466. https://doi.org/10.1038/d41586-022-01480-x

Das D (2017) Park, people and biodiversity conservation in Kaziranga National Park, India. Space Cult India 5(1):36–48. https://doi.org/10.20896/saci.v5i1.244

Dorji T, Yangzom D, Norbu N, Rinchen S, Dorjee J, Tenzin T (2023) Understanding the impact of climate change and resilience among highlanders in northern parts of Bhutan: a case study in Gasa district. PLOS Clim 2(4):e0000079. https://doi.org/10.1371/journal.pclm.0000079

Easton-Gomez SC, Mouritz M, Breadsell JK (2022) Enhancing emotional resilience in the face of climate change adversity: a systematic literature review. Sustainability (Switzerland) 14(21). https://doi.org/10.3390/su142113966

Ecosystem-based adaptation—stakeholder dialogue report. https://doi.org/10.59117/20.500.11822/42198

Falk DA, Van Mantgem PJ, Keeley JE, Gregg RM, Guiterman CH, Tepley AJ, Young DJN, Marshall LA (2022) Mechanisms of forest resilience. For Ecol Manage 512:120129. https://doi.org/10.1016/j.foreco.2022.120129

FAO (2016) Lessons learned for decision makers from a review of experiences of the terrafrica strategic investment programme on SLM in sub-Saharan Africa (SIP) under the nepad-terrafrica partnership framework. In: Informing future interventions for scaling-up sustainable land management

Fisher B, Turner RK, Morling P (2009) Defining and classifying ecosystem services for decision making. Ecol Econ 68(3):643–653. https://doi.org/10.1016/j.ecolecon.2008.09.014

Folke C, Carpenter SR, Walker B, Scheffer M, Chapin T, Rockström J (2010) Resilience thinking: integrating resilience, adaptability and transformability. Ecol Soc 15(4):20. https://doi.org/10.5751/ES-03610-150420

Gogoi N, Sarma A, Choudhury M, Samanta P, Faizan M, Sadhak S (2024a) Soil heavy metal pollution and ecological risk assessment in disturbed and undisturbed soil of Morigaon, Assam. Discover Environ 2(1):13. https://doi.org/10.1007/s44274-024-00039-z

Gogoi N, Choudhury M, Ul Hasan MS, Changmai B, Baruah D, Samanta P (2024) Soil organic carbon pool in diverse land utilization patterns in North-East India: an implication for carbon sequestration. Environ Dev Sustain, 1–27. https://doi.org/10.1007/s10668-024-05037-y

Gunderson LH (2002) Ecological resilience—in theory and application. Annu Rev Ecol Evol Syst 31(1):425–439. https://doi.org/10.1146/annurev.ecolsys.31.1.425

Gunderson LH, Holling CS (2002) Panarchy: understanding transformations in human and natural systems. Island Press, Washington

Hayes S, Desha C, Burke M, Gibbs M, Chester M (2019) Leveraging socio-ecological resilience theory to build climate resilience in transport infrastructure. Transp Rev 39(5):677–699. https://doi.org/10.1080/01441647.2019.1612480

Intergovernmental Panel on Climate Change (2022) Climate change 2022: impacts, adaptation and vulnerability. Cambridge University Press, Cambridge. https://doi.org/10.1017/9781009325844

Karsai I, Schmickl T, Kampis G (2020) Understanding ecosystem stability and resilience through mathematical modeling. In: Resilience and stability of ecological and social systems. Springer, Cham, pp 1–17. https://link.springer.com/chapter/10.1007/978-3-030-54560-4_1

Khan MSI (2021) Supporting ecosystem services: concepts and linkages to sustainability. In: Leal Filho W, Azul AM, Brandli L, Lange Salvia A, Wall T (eds) Encyclopedia of the UN sustainable development goals: life on land. Springer, Cham, pp 969–987. https://doi.org/10.1007/978-3-319-95981-8_34

Koellner T (2011) Ecosystem services and global trade of natural resources. Routledge, London

Kumar D, Choudhury M (2021) Recognizing economic values of wetland ecosystem services: a study of emerging role of monetary evaluation of Chandubi ecosystem and biodiversity. In: Wetlands conservation: current challenges and future strategies, pp 87–110. https://doi.org/10.1002/9781119692621.ch5

Landa-Ochoa JL, Gallardo-López F, Escamilla-Prado E, Cerdán-Cabrera CR, Ortiz-Ceballos GC (2024) Social-ecological resilience: a key concept for the study of sustainability in agroecosystems. Trop Subtrop Agroecosyst 27(1). https://doi.org/10.56369/tsaes.4758

Leão TC, Lobo D, Scotson L (2017) Economic and biological conditions influence the sustainability of harvest of wild animals and plants in developing countries. Ecol Econ 140:14–21. https://doi.org/10.1016/j.ecolecon.2017.04.030

Levin SA (1999) Fragile dominion: complexity and the commons. Perseus Books, Cambridge, Massachusetts.

Manning P, Van Der Plas F, Soliveres S, Allan E, Maestre FT, Mace G, Whittingham MJ, Fischer M (2018) Redefining ecosystem multifunctionality. Nat Ecol Evol 2(3):427–436. https://doi.org/10.1038/s41559-017-0461-7

Mngumi LE (2020) Ecosystem services potential for climate change resilience in peri-urban areas in Sub-Saharan Africa. Landscape Ecol Eng 16(2):187–198. https://doi.org/10.1007/s11355-020-00411-0

Narura AK, Ssenku JE, Kudamba A, Okurut SA, Namuli A, Nansikombi N, Nakizito J, Kafeero HM, Walusansa A (2024) The environmental conservation strategies used in secondary schools align with guidelines of united nations environmental program: evidence from a cross-sectional study in a city setting. Asian J Environ Ecol 23(7):200–230. https://doi.org/10.9734/ajee/2024/v23i7575

Nelson E, Mendoza G, Regetz J, Polasky S, Tallis H, Cameron D, Chan KM, Daily GC, Goldstein J, Kareiva PM, Lonsdorf E, Naidoo R, Ricketts TH, Shaw M (2009) Modeling multiple ecosystem services, biodiversity conservation, commodity production, and tradeoffs at landscape scales. Front Ecol Environ 7(1):4–11. https://doi.org/10.1890/080023

Ossewaarde R, Filatova T, Georgiadou Y, Hartmann A, Ozerol G, Pfeffer K, Stegmaier P, Torenvlied R, Van Der Voort M, Warmink J, Borsje B (2021) Review article: towards a context-driven research: a state-of-the-art review of resilience research on climate change. Nat Hazard 21(3):1119–1133. https://doi.org/10.5194/nhess-21-1119-2021

Paulauskaite-Taraseviciene A, Lagzdinyte-Budnike I, Gaiziuniene L, Sukacke V, Daniuseviciute-Brazaite L (2022) Assessing education for sustainable development in engineering study programs: a case of AI ecosystem creation. Sustainability (Switzerland) 14(3):1–22. https://doi.org/10.3390/su14031702

Potschin M, Haines-Young R (2016) Defining and measuring ecosystem services. In: Potschin M, Haines-Young R, Fish R, Turner RK (eds) Routledge handbook of ecosystem services. Routledge, Taylor and Francis Group, London, pp 25–44

Ryplova R, Pokorny J, Baxa M (2023) Education for sustainability: innovative teaching on photosynthesis of aquatic plants in ecological context. Eur J Sustain Dev 12(4):69. https://doi.org/10.14207/ejsd.2023.v12n4p69

Santos E, Carvalho M, Martins S (2023) Sustainable water management: understanding the socioeconomic and cultural dimensions. Sustainability 15(17):13074

Scharlemann Hasan N, Choudhary S, Naaz N, Sharma N, Laskar RA (2021) Recent advancements in molecular marker-assisted selection and applications in plant breeding programmes. J Genet Eng Biotechnol 19(1). https://doi.org/10.1186/s43141-021-00231-1

Schneider T, Ruby Leung L, Wills RCJ (2024) Opinion: optimizing climate models with process knowledge, resolution, and artificial intelligence. Atmos Chem Phys 24(12):7041–7062. https://doi.org/10.5194/acp-24-7041-2024

Schultz L, Folke C, Österblom H, Olsson P (2015) Adaptive governance, ecosystem management, and natural capital. Proc Natl Acad Sci 112(24):7369–7374

Singletary L, Koebele E, Evans W, Copp CJ, Hockaday S, Rego JJ (2022) Evaluating stakeholder engagement in collaborative research: co-producing knowledge for climate resilience. Socio-Ecol Pract Res 4(3):235–249. https://doi.org/10.1007/s42532-022-00124-8

Stephenson TS, Sörensson AA, Castellanos E, Ley D, Mahon R (2014) Challenges for climate change adaptation in Latin America and the Caribbean region. Front clim 6:1392033. https://doi.org/10.3389/fclim.2024.1392033

Strassburg BBN, Iribarrem A, Beyer HL, Cordeiro CL, Crouzeilles R, Jakovac CC, Braga Junqueira A, Lacerda E, Latawiec AE, Balmford A, Brooks TM, Butchart SHM, Chazdon RL, Erb KH, Brancalion P, Buchanan G, Cooper D, Díaz S, Donald PF, Kapos V, Leclère D, Miles L, Obersteiner M, Plutzar C, de M. Scaramuzza CA, Scarano FR, Visconti P (2020) Global priority areas for ecosystem restoration. Nature 586(7831):724–729. https://doi.org/10.1038/s41586-020-2784-9

Tang X, Adesina JA (2022) Biodiversity conservation of national parks and nature-protected areas in West Africa: the case of Kainji National Park, Nigeria. Sustainability (Switzerland) 14(12). https://doi.org/10.3390/su14127322

Tanguay L, Herzog LM, Audet R, Beisner BE, Martin R, Pahl-Wostl C (2023) Opportunities for and barriers to anticipatory governance of two lake social–ecological systems in Germany and Canada. People Nat 5(3):911–928. https://doi.org/10.1002/pan3.10464

Tanner T, Bahadur A, Moench M (2017) Challenges for resilience policy and practice

Tripathy DAB, Swain DBC, Mishra MS (2024) Environmental sustainability for a sustainable future and role of education (in climate change perspectives). Educ Adm Theory Pract 30(5):13660–13665. https://doi.org/10.53555/kuey.v30i5.5952

Tuan H (2024) Agricultural practices and climate resilience: case study in Vietnam. Int J Clim Stud 3(1):24–36. https://doi.org/10.47604/ijcs.2476

Wang F, Wong WK, Wang Z, Albasher G, Alsultan N, Fatemah A (2023) Emerging pathways to sustainable economic development: an interdisciplinary exploration of resource efficiency, technological innovation, and ecosystem resilience in resource-rich regions. Resour Policy 85:103747

Zaveri P, Patel R, Munshi NS (2020) Trickling of itinerant nanoparticles in wastewater effluents, pp 1–21. https://doi.org/10.1007/978-3-030-38192-9_1

Zellmer AJ, Goto BS (2022) Urban wildlife corridors: Building bridges for wildlife and people. Front Sustain Cities. https://doi.org/10.3389/frsc.2022.954089

Zickgraf C, Ali SH, Clifford M, Djalante R, Kniveton D, Brown O, Ayeb-Karlsson S (2022) Natural resources, human mobility and sustainability: a review and research gap analysis. Sustain Sci 17(3):1077–1089. https://doi.org/10.1007/s11625-021-01073-z

Part II
Greening Our Future: Realizing Climate Resilience Through Ecosystems

Chapter 5
Nature's Comeback: Case Studies of South Asian Countries Regarding Biodiversity Restoration and Conservation

Shumaila Batool, Murad Muhammad, Minahal Fatima, Khalid Abbas, Kehu Li, Fatima Amin, and Muqadas Batool

Abstract Biodiversity encompasses the vast array of living organisms in a habitat, crucial for ecosystem functioning and human health. Humans derive advantages from biodiversity across environmental, economic, and cultural dimensions. Biodiversity provides various ecosystem services such as climate regulation, pollutant control, soil erosion prevention, conservation of natural habitats, and crop pollination. Biodiversity is decreasing at a faster rate in the current situation. Several studies have demonstrated a decrease in biodiversity and the extinction of numerous species. Human actions such as overexploitation, pollution, global warming, and habitat loss are the primary causes of biodiversity decline. Restoring and conserving biodiversity is crucial for maintaining environmental health and human well-being. It is imperative to preserve essential animal and plant species. Various methods, including in-situ and

---

S. Batool (✉) · M. Fatima · K. Abbas · F. Amin
Department of Zoology, Wildlife and Fisheries, University Of Agriculture, Faisalabad 38000, Pakistan
e-mail: Shumailabatool262@gmail.com

M. Muhammad (✉)
State Key Laboratory of Desert and Oasis Ecology, Key Laboratory of Ecological Safety and Sustainable Development in Arid Lands, Xinjiang Institute of Ecology and Geography, Chinese Academy of Sciences, Beijing, China
e-mail: muradbotany1@uop.edu.pk

University of Chinese Academy of Sciences, Beijing 100049, China

Xinjiang Key Laboratory of Biodiversity Conservation and Application in Arid Lands, Xinjiang Institute of Ecology and Geography, Chinese Academy of Sciences, Beijing, China

K. Li
College of Life Sciences, Guizhou University, Guiyang 550025, Guizhou, China

M. Batool
Department of Botany, (Govt Graduate College Layyah) Bahauddine Zakariya University, Multan 60800, Pakistan

M. Choudhury et al. (eds.), *Sustainable Synergy: Harnessing Ecosystems for Climate Resilience*, Climate Change Management,
https://doi.org/10.1007/978-3-031-77957-2_5

ex-situ conservation, are employed to restore and protect biodiversity. In-situ conservation utilizes strategies like national parks, biosphere reserves, and wildlife sanctuary regions. Ex-situ conservation involves botanical gardens, zoological gardens, aquaria, and in-vitro processes. Hence, preserving biodiversity is crucial to prevent species extinction and protect endangered and indigenous species. This chapter identifies the primary risks to biological diversity resulting from human activities and examines various conservation and restoration techniques for biodiversity.

Keywords Biodiversity · Threats · Conservation · Restoration · Habitat loss

Introduction

Biodiversity conservation entails preserving the biodiversity's life on the planet while ensuring the proper functioning and survival of ecological systems. This encompasses the preservation, maintenance, efficient utilization, restoration, and improvement of the many elements of biodiversity. Biodiversity conservation is becoming increasingly recognized as a crucial element of sustainability (Maiti and Maiti 2023). Biodiversity, the collective term for the vast array of biological diversity on Earth, results from the interaction between organisms and their surroundings and associated ecological processes. This includes microbes, plants, and animals. Biological diversity is essential for sustaining the ability of ecosystems to resist and recover from disturbances, as well as for supplying valuable benefits to ecosystems and sustaining human livelihoods. The significance of biodiversity is incalculable since it involves several essential functions such as medicine, pollination, beauty, and nutrition cycling (Kumar and Choudhury 2021). An ecosystem's wide range of animals forms a sustainable foundation every community or country uses for its progress, advancement, and improvement (Macdonald 2023).

In the current circumstances, the rate at which biodiversity is declining and the changes occurring in the environment are at a record high, and there is no indication that this process will slow down any time soon. Due to human activity, almost all of Earth's ecosystems have undergone substantial disruption and alteration, and they are constantly being modified for agriculture and other purposes. The number and geographical distribution of many species of plants and animals have decreased. Species extinction is a natural process in the planet's history, but anthropogenic activity has significantly accelerated the extinction rates, exceeding its standard rate by at least 100 times (Shivanna 2020). Various factors contribute to the decline of biological diversity, including natural and anthropogenic elements that directly or indirectly disrupt ecosystem functions.

The reduction in biodiversity has significant environmental and socio-economic consequences that affect food availability, ecosystem functioning, human well-being, and the existence of other organisms (Gogoi et al. 2024a, b). To create efficient restoration and conservation techniques, it is essential to comprehend the fundamental reasons for declining biodiversity. Several causes are contributing to the fast

Table 5.1 Major threats to biodiversity worldwide

Major threats	Descriptions	References
Habitat degradation	Habitat degradation is the leading cause of declining biodiversity globally. Degradation makes whole environments unsuitable for their species. Displacement or destruction of habitat animals reduces biodiversity	Okorondu et al. (2022)
Climate change	Elevated temperatures, altered precipitation patterns, and severe weather phenomena disturb the balance of ecosystems and habitats, leading to changes in species' geographic range, decreases in populations, and the disappearance of specialized habitats	Siddha and Sahu (2022) Haldar et al. (2023)
Overexploitation	Overexploitation occurs when individuals of a particular species are collected at a pace that exceeds the population's natural reproductive capability. These activities may be accomplished by hunting, fishing, trading, and gathering food. Overexploitation poses a significant risk to several species, including marine fish and invertebrates, plants, and animals targeted for meat consumption	Hughes et al. (2023)
Pollution	Pollution originating from many sources, such as factories and urban regions, contaminates the environment, lakes, rivers, and soil, resulting in the decline of biodiversity and detrimental effects on the health of the ecosystem	Singh et al. (2020)
Disease	Diseases, both naturally occurring and brought about by humans, can destroy animal populations, particularly when those species are immune-compromised or have low genetic variations	Yadav and Upadhyay (2023)
Invasive species	Introducing exotic species into habitats may result in the displacement of native species, transmission of diseases, and disruption of biological functions, ultimately causing a reduction in biological diversity	Siddiqui et al. (2021)

decline in biodiversity, including changes in land use, which modify and destroy habitats, the over-exploitation of biological resources, changes in the climate, pollution, and invasive species (Davison et al. 2021), as shown in (Table 5.1).

Principles of Biodiversity Restoration and Conservation

The primary objective of most restoration initiatives is the preservation of biodiversity, which includes activities such as reintroducing endangered plants and animal species as well as restoring a wide variety of species. Restoration is the deliberate manipulation of species and ecological functions to self-regulating, persistent, indigenous environments seamlessly integrated into the natural environment, resembling their original state before human-induced disruptions as closely as feasible. The restoration of biodiversity ensures the preservation of essential ecological diversity, which is necessary for maintaining the continuous functioning of food webs. The conservation of animals and their surrounding ecosystems is necessary to mitigate the destructive actions caused by humanity throughout history (Geist and Hawkins 2016).

Biodiversity restoration and conservation strategies include various actions that attempt to preserve and restore multiple species and habitats. These techniques, such as integration, connectivity, and representativeness, aim to improve biological diversity by concentrating on restoring habitats in fractured environments. Furthermore, restoring habitats that have been destroyed and are home to historical biodiversity remnants is paramount. This restoration process involves using GIS analysis, fieldwork, and phytosociological tools to enhance the ability of the landscape to withstand and recover from disturbances (Del Tánago et al. 2021). To successfully restore biodiversity, it is crucial to re-establish a wide range of genetic variations and species, pay attention to specific locations for conservation, repair ecosystems and selective plantations, and consistently maintain the landscape for long-term viability. It is crucial to facilitate connection among various ecosystem components since it enables species migration and facilitates genetic interchange. Preserving biodiversity and ecosystem services requires implementing essential techniques such as establishing protected areas and promoting plant variety in restoration efforts. Continual learning and adaptation are critical to the repair process as professionals modify their approaches in response to feedback and monitoring data. Engaging with the local people is crucial since they often contain important information about the region and may provide insights regarding conservation strategies. Restoration professionals may achieve successful ecological recovery, excellent plant and animal diversity, and enhanced resilience to climate change by adhering to these principles (Bennett et al. 2017).

Restoration and Conservation Strategies

Habitat Restoration Techniques

The UN defines habitat restoration as "preventing and reversing deterioration, which results in recovered ecosystem functions and restored biological diversity" (Abhilash 2021). The significance of these recovered habitats lies not only in their aesthetic appeal and biological diversity but also in their crucial role in providing essential ecosystem services such as purification of air and water, flood mitigation, fertility of soil, and availability of natural resources. Restoring habitats is critical for mitigating the adverse effects caused by human activities on our Planet. Restoring the regions that have experienced habitat loss is crucial since it is a key element contributing to biodiversity extinction. Environmental destruction significantly affects the survival and number of flora and fauna, the accessibility of natural assets, the condition of our soil, water, and air, and the overall quality of life for all inhabitants of the

Planet. The United Nations Environment Programme (UNEP) states that environmental destruction negatively impacts the well-being of around 3.2 billion individuals, which accounts for 40% of the worldwide population (Brusseau et al. 2019). Restoring habitat plays a crucial role in avoiding a breakdown of the food chain. It is predicted that by effectively restoring 15% of degraded areas, we may prevent 60% of the anticipated extinctions of species (Arneth et al. 2021).

Reforestation and Wetland Restoration

Reforestation and wetland restoration are essential to recovering damaged land and preserving biological diversity. Reforestation is the deliberate act of planting indigenous plants in regions that have experienced deforestation to enhance the ability of ecosystems to withstand disturbances and the absorption of carbon dioxide (Gogoi et al. 2024a, b). Essential actions include choosing appropriate locations and species, the use of different techniques for planting, and implementing continuous care and monitoring. Wetland restoration aims to rehabilitate damaged wetland habitats to improve their ecological purpose and promote the presence of various flora and fauna species (Choudhury et al. 2021). This often involves the restoration of the natural water cycle, the reintroduction of native plants, the management of exotic species, and the improvement of habitat characteristics (Kumar et al. 2020). Monitoring is essential for evaluating the status of restoration efforts and providing guidance for adaptive management techniques. Reforestation and wetland restoration contribute to the connectivity of habitats, preservation of watersheds, and reduction of global warming. This makes them valuable instruments for conserving biodiversity and restoring ecosystems (Cai et al. 2021).

Species Reintroduction Programs and Captive Breeding Efforts

Captive breeding and reintroduction are crucial techniques in the conservation of threatened species. These techniques aim to enhance the number of individuals and biodiversity by fostering animals in controlled circumstances and restoring them to their original habitats. In many cases, human involvement is necessary to save endangered species at risk of extinction because of small populations. Humans may assist by implementing captive breeding and reintroduction programs. Reintroduction programs are a kind of ex-situ conservation technique in which individuals of a species at risk of extinction are placed in captivity to breed offspring with the help of humans. Reintroduction programs are often implemented inside the confines of zoos and sanctuaries. Captive breeding is especially crucial for species experiencing ongoing loss in their natural habitats and lacking robust populations in their native environment. Captive breeding for release may also serve as a means to enhance and blend the genetic variety obtained from the capture of animals in the wild before their release. The ultimate objective of these programs is to reintroduce captive-bred animals into their original habitats and support the establishment of self-sustaining populations in the wild. A self-sustaining population can survive without human interference (Harley et al. 2018).

The Middle Eastern antelope species known as the Arabian oryx is one of the earliest success stories of reintroduction efforts (Elmeer et al. 2019). The captive breeding program was initiated in 1962 when the species became extinct in its natural habitat. The most recent remaining individuals were captured, and their descendants were first reintroduced to Oman in 1982. Their descendants were later found in Jordan, the United Arab Emirates, Saudi Arabia, and Israel (Ye 2017). The species is currently classified as vulnerable according to the IUCN criteria, which denotes a reduced risk compared to being endangered. Moreover, evidence suggests that natural populations show signs of stabilization over time. The captive breeding of California condors, blue tangs, and black-footed ferrets are among the other noteworthy tales (Madliger et al. 2020).

Landscape-Scale Conservation Approaches and Corridor Connectivity

Landscape-scale conservation is a technique that considers and maintains the biological, social, and economic components of a landscape or geographic region. It includes protecting and managing biological systems and landscapes across extensive geographical areas, usually including several governing bodies and property ownerships. In the last 20 years, maintaining landscape connectedness has become critical for preserving plant and animal life, maintaining healthy environments and animal populations, and facilitating wildlife species in adapting to changing climates (Kremen and Merenlender 2018). The primary strategy for establishing connectedness in fragmented environments is the establishment of habitat corridors.

The decrease in landscape connection has detrimental implications on biodiversity globally since landscape connectivity plays an essential role in species survival because of changing events like recolonization, seasonal movement, and dispersal. Therefore, one of the most significant difficulties for biodiversity conservation is maintaining connectedness, especially in places where agricultural lands and transportation infrastructure are rapidly encroaching into and displacing natural ecosystems. Efficient corridors that are ecologically and functionally effective are considered to be conservation measures that promote landscape connectivity. These corridors have been extensively studied and discussed in the literature. Efficiently designed and effectively controlled corridors may enhance the movement of species and promote essential processes like gene exchange and natural regeneration. Consequently, this can positively impact the preservation of biodiversity at both local and regional levels (Hilty et al. 2019).

Community Engagement and Stakeholder Involvement

Importance of Engaging Local Communities and Stakeholders in Conservation and Restoration Projects

The depletion of biodiversity caused by anthropogenic actions is a significant concern in pursuing environmental sustainability. Conserving ecological systems is now widely recognized as an ethical obligation for decision-makers, particularly in companies dependent on natural resources. Companies must exhibit dedication to biodiversity concerns and provide stakeholders additional details about their actions. The active involvement of local stakeholders is a fundamental aspect of several biodiversity preservation and natural resource management initiatives worldwide. Biodiversity projects often include four primary categories of stakeholders: non-governmental organizations (NGOs), public bodies, specialists and educational institutions, and collaborations of businesses (Raimi et al. 2022).

Stakeholder participation plays a crucial role in biodiversity projects and offers advantages corporations would have needed to have attained independently. Many vital actions, such as establishing new standards and certifying them, conducting research initiatives and evaluating their effect, conducting biodiversity surveys, and creating protected areas, rely heavily on the involvement of stakeholders. These measurements are distinguished by their complicated nature and prominent societal presence (Boiral and Heras-Saizarbitoria 2017). Implementing significant actions, such as establishing protected areas and adopting certifiable standards, also enhances the public's acceptance of companies regarding ecologically essential matters. This discovery aligns with neo-institutional theory, which emphasizes the influence of external forces and the business community's goal of social legitimacy in implementing new techniques.

The past few decades have seen institutional demands to save biodiversity rise. Based on a survey of 1043 managers, over 50% of participants noticed diversity as crucial for establishing, sustaining, or enhancing a company's reputation (Boiral and Heras-Saizarbitoria 2017). According to neo-institutionalism theories, implementing innovative methods, particularly in managing the environment, is driven by outside forces and the achievement of company legitimacy. Engaging stakeholders in biodiversity management may enhance a company's credibility and reduce external pressures, particularly for enterprises that extract natural resources for profit. Engaging stakeholders may improve the efficacy of managing biodiversity (Chedrawi et al. 2020).

Incorporating Traditional Ecological Knowledge and Indigenous Perspectives

Indigenous peoples actively collaborate as partners in preserving biodiversity, which thrives in local regions. They possess extensive knowledge about the behavior of intricate ecosystems in their specific areas and have a long-standing history of using resources in a certain way. Traditional ecological knowledge (TEK) is essential for the preservation and accountable utilization of biological diversity. Biodiversity conservation encompasses indigenous communities' perspectives and spiritual beliefs, which contribute to developing novel environmental ethics and traditional methods of managing natural resources that have been tried and proven over several generations (Dkhar and Tiwari 2020). TEK plays a crucial role in biodiversity conservation across various levels. This includes the conventional understanding of fauna, flora, soils, and landscapes, which helps ensure the sustainable utilization of assets.

Additionally, TEK involves social institutions or organizations that facilitate coordination, cooperation, rule-making, and rule enforcement. Lastly, TEK contributes to ecological perception and provides significance to social interactions. Furthermore, TEK has significant implications for ethnoecology, population ecology, species interaction, and forest management. It is suggested that biological diversity managers and Western scientists build close relationships with knowledge holders, learn each other's communication styles, build trust to work together, and find ways to benefit from each other's expertise to conserve biodiversity (Albuquerque et al. 2021).

Monitoring and Evaluation

Importance of Monitoring Biodiversity Restoration Outcomes

Keeping biodiversity under surveillance is essential. Monitoring can be done in various ways. The Utmost way to monitor wildlife is through their long-term patterns (habitat quality, species distribution, and population dynamics). It can also be done by adequately assessing the productivity of restoring strategies and adjudicating the different scenarios in contracting the pre- and post-construction to determine which is best for the lookout. As it concludes which is better, the plan is modified per demand (Crandall et al. 2023).

Techniques for Assessing Ecosystem Recovery and Measuring Success

Surveys are among the most significant ways to assess recovery in the fauna—surveys of the habitat, surveys of the wildlife, and last but not least, remote sensing. Genetic Monitoring is also an innovation that checks wildlife recovery. Genetic Monitoring includes the Genetic Modification of Organisms and DNA Sequences. This is done to store the hereditary material of rare or endangered species. The biodiversity and habitat assessment surveys can be done using satellite images (remote sensing) to check whether the species prosper. This survey was also conducted to check the evolutionary processes over time. Habitat fragmentation is done to check the quality and moisture content of the soil. Checking the soil is pivotal as it administrates the ecosystem's health (Kerry et al. 2022).

Long-Term Monitoring Programs and Adaptive Management Strategies

The pivotal biodiversity assessment can be done using various programs' strategies. One of the Utmost programs is the Establishing a Standardized Monitoring Protocol (SMP), whereas the other programs also have significance. Some of the strategies are as shown in Table 5.2.

Table 5.2 Monitoring programs and adaptive management strategies

Programs	Strategies	References
Establishing a standardized monitoring protocol (SMP)	Creating an SMP allows for comparing outcomes over time and guarantees consistency in data collection across many sites	Logue and Grimes (2022)
Long-term data collection	Long-term programs enable the identification of patterns and trends that might not be noticeable in short-term evaluation programs, leading to more comprehensive dynamics of the ecosystem	Lindenmayer et al. (2022)
Adaptative management	Data collected in the Long-Term Monitoring Data (LTMD), adaptive Management allows restoration tactics to be literately modified in response to shifting circumstances, previous ones, and fresh knowledge	Van Wilgen et al. (2022)
Stakeholder collaboration	A diverse committee, including scientists, land managers, local population, and policymakers, is essential in tracking and assessing procedures to improve data collection and raise the applicability of findings for making decisions	Jaya et al. (2022)

Fig. 5.1 Giant Panda (*Ailuropoda melanoleuca*) (Wei et al. 2020)

Successful Case Studies in Biodiversity Conservation

Certain case studies have led to the successful conservation of biodiversity. Some of them are as follows.

Giant Panda

The Giant Pandas (*Ailuropoda melanoleuca*) are commonly found in China. Most Pandas, especially the Giant Pandas (Fig. 5.1), are becoming endangered because of habitat loss and fragmentation. One of the main reasons for habitat loss and fragmentation is the development and expansion of industrial areas. They all lived in 33 different regions because of the mountain's presence, and the other reason is humans; that's why they all get different habitats in which they cannot adjust and start to die. After that, the Chinese government took action and created the Protected Areas (PA) to conserve the species of *A. melanoleuca;* now, they are growing in numbers (Kang 2022).

Madagascar's Masoala National Park

The National Park of Madagascar was created in 1997. This park has a unique forestation pattern and an extreme diversity of marine life as it connects the park with the sea. The destruction of the flora was seen due to anthropogenic and non-anthropogenic forest degradation (Fig. 5.2). To save the flora, biodiversity conservation was done. Not only this, the locals also cut down the trees for the sake of precious wood (Ebony, Rosewood, *Dalbergia* spp.) Conservationists of Masoala Park and the organizations concerned and connected with the park decided to take the initiative and start to work on preserving the flora. To restrain the anthropogenic and non-anthropogenic factors, the crops (rice) and the diverse species of herbs and shrubs were transferred

Fig. 5.2 Anthropogenic and non-anthropogenic factors (Burivalova et al. 2015)

to another place (far away from steep slopes), and strict laws against the smuggling of wood (Burivalova et al. 2015).

Whitetip Shark

The Epipelagic Species Oceanic Whitetip Shark (*Carcharhinus longimanus*) is mainly found in tropical and sub-tropical water. The main reason that the *C. longimanus* (Fig. 5.3) is an endangered species is over-exploitation. Most of the bycatches accidentally catch the Shark, which mixes it with the other shark species. Sharks get captured by the bycatches because they remain on the top layer of the water's surface. Over time, the exploitation of the Sharks got worse, and then actions were taken by the International Commission for the Conservation of Atlantic Tunas (ICCAT), Western and Central Pacific Fisheries Commission (WCPFC), and Indian Ocean Tuna Commission (IOTC). They started the work on preserving Sharks and applied some rules and regulations that are obligatory for everyone. Their exploitation is gradually decreasing, but in some areas, the locals have demolished the rules (Young and Carlson 2020).

Fig. 5.3 Oceanic Whitetip Shark (*Carcharhinus longimanus*) (Young and Carlson 2020)

Bengal Tigers

The Bengal Tiger (*Panthera tigris tigris*) is mainly found in the Sundarban Mangrove Forest, Bangladesh. This species is found in Bangladesh, India, and Nepal. The main reason that makes the *P. tigris tigris* an endangered species is human threats. The mainland of the tigers is the area of Sundarbans Forest, but human interference makes the tigers' survival stricter day after day, and the tigers dominate the land of 10,262 km^2. However, nowadays, they are all reserved for the sanctuaries and the reserve area of the forest just because of fragmentation and habitat loss (Islam 2023). Their population declined gradually, and steps were taken to conserve it. The Global Tiger Initiative program was held in 2010 with the urge to double the number of tigers by 2022. Only two countries take the initiative seriously, Nepal and India; due to this, the number of tigers is growing there. In Bangladesh, the Wildlife Team was created to protect the tigers from hunting, and in Nepal, the Banke National Park was established with a surface area of 210 sq. mi. with the vow to double the number of tigers (Shrestha et al. 2021).

Sri Lankan Elephants

The subspecies of Asian Elephant is the Sri Lankan Elephant (*Elephas maximus maximus*) mainly found in Sri Lanka. The main through which these species become endangered is Human- Human-elephant conflict. This conflict is majorly for land as the human population increases gradually there was a viable decline in the population of the elephant due to a human development program that converted the natural area to irrigated land (Prakash et al. 2020). To prevent the Elephants from extinction the Department of Wildlife take the initiative. To conserve them, the Udawalawe Elephant Transit Center in Udawalawe National Park was created where the elephants are free to go wherever they want without any restriction (Santiapillai et al.).

Table 5.3 Challenges and opportunities (Hoffmann 2022)

Challenges	Opportunities
Human land use	People and bio-culture-centered conservation
Climate change	Climate-smart conservation
Human conflicts	People-centered conservation
Uncertain future development	Agile governance
Insufficient data and monitoring	Use satellite pictures to keep wildlife under surveillance

Challenges and Opportunities of Area-Based Conservation

Wildlife faces specific challenges daily. Some difficulties comprise human land use, which occupies 75% of the land. Inducement in climate also produces uncertainty in the susceptibility of the species. As the environment shows the variation in the temperature, it enhances the chances of acidification (acid rain), which is an alarming situation for the survival of the fauna. Specific strategies have been employed and named as opportunities to prevent wildlife from being threatened and endangered. Opportunities are mainly comprised of people and bio-culture-centered conservation (Hoffmann 2022). Some of the challenges and opportunities are as follows, Table 5.3.

Conclusion

Earth's ecosystems and the many species they support depend on conservation and biodiversity restoration initiatives. Biodiversity has a significant role in the existence of the ecosystem, including humans, as it plays the foremost role in food security, the cure of illness, and the increase in economic growth. It takes an integrated strategy to solve problems like pollution, overexploitation, invasive species, habitat loss, and climate change. Specific strategies are involved in restoring wildlife, but some exceptional strategies include reforestation and wetlands. Stakeholder involvement and evidence-based decision-making are crucial for successful conservation efforts, and monitoring and assessment are critical for evaluating progress and directing adaptive management plans. Cooperation and creativity are essential to achieving conservation goals and incorporating cutting-edge technologies.

References

Abhilash PC (2021) Restoring the unrestored: strategies for restoring global land during the UN decade on ecosystem restoration (UN-DER). Land 10:201

Albuquerque UP, Ludwig D, Feitosa IS, de Moura JMB, Gonçalves PHS, da Silva RH, da Silva TC, Gonçalves-Souza T, Ferreira Junior WS (2021) Integrating traditional ecological knowledge into academic research at local and global scales. Reg Environ Change 21:1–11

Arneth A, Olsson L, Cowie A, Erb K-H, Hurlbert M, Kurz WA, Mirzabaev A, Rounsevell MD (2021) Restoring degraded lands. Annu Rev Environ Resour 46:569–599

Bennett NJ, Roth R, Klain SC, Chan K, Christie P, Clark DA, Cullman G, Curran D, Durbin TJ, Epstein G (2017) Conservation social science: understanding and integrating human dimensions to improve conservation. Biol Conserv 205:93–108

Boiral O, Heras-Saizarbitoria I (2017) Managing biodiversity through stakeholder involvement: why, who, and for what initiatives? J Bus Ethics 140:403–421

Brusseau M, Ramirez-Andreotta M, Pepper I, Maximillian J (2019) Environmental impacts on human health and well-being. Environ Pollut Sci, 477–499

Burivalova Z, Bauert MR, Hassold S, Fatroandrianjafinonjasolomiovazo NT, Koh LP (2015) Relevance of global forest change data set to local conservation: case study of forest degradation in Masoala National Park, Madagascar. Biotropica 47:267–274

Cai Y, Liang J, Zhang P, Wang Q, Wu Y, Ding Y, Wang H, Fu C, Sun J (2021) Review on strategies of close-to-natural wetland restoration and a brief case plan for a typical wetland in Northern China. Chemosphere 285:131534

Chedrawi C, Osta A, Osta S (2020) CSR in the Lebanese banking sector: a neo-institutional approach to stakeholders' legitimacy. J Asia Bus Stud 14:143–157

Choudhury M, Sharma A, Singh P, Kumar D (2021) Impact of climate change on wetlands, concerning Son Beel, the largest wetland of North East, India. In: Global climate change. Elsevier, pp 393–414. https://doi.org/10.1016/B978-0-12-822928-6.00006-X

Crandall ED, Toczydlowski RH, Liggins L, Holmes AE, Ghoojaei M, Gaither MR, Wham BE, Pritt AL, Noble C, Anderson TJ (2023) Importance of timely metadata curation to the global surveillance of genetic diversity. Conserv Biol 37:e14061

Davison CW, Rahbek C, Morueta-Holme N (2021) Land-use change and biodiversity: challenges for assembling evidence on the greatest threat to nature. Glob Change Biol 27:5414–5429

Del Tánago MG, Martínez-Fernández V, Aguiar FC, Bertoldi W, Dufour S, de Jalón DG, Garófano-Gómez V, Mandzukovski D, Rodríguez-González PM (2021) Improving river hydromorphological assessment through better integration of riparian vegetation: scientific evidence and guidelines. J Environ Manage 292:112730

Dkhar M, Tiwari B (2020) Traditional ecological knowledge of tribal communities of North East India. Biodiversitas J Biol Divers 21

Elmeer K, Matatt I, Al-Malki A, Alam MS (2019) Genetic diversity and relatedness in Arabian oryx (*Oryx leucoryx*) revealed by SSR markers. Qscience Connect 2019:3

Geist J, Hawkins SJ (2016) Habitat recovery and restoration in aquatic ecosystems: current progress and future challenges. Aquat Conserv Mar Freshwat Ecosyst 26:942–962

Gogoi N, Sarma A, Choudhury M, Samanta P, Faizan M, Sadhak S (2024a) Soil heavy metal pollution and ecological risk assessment in disturbed and undisturbed soil of Morigaon, Assam. Discover Environ 2(1):13. https://doi.org/10.1007/s44274-024-00039-z

Gogoi N, Choudhury M, Ul Hasan MS, Changmai B, Baruah D, Samanta P (2024) Soil organic carbon pool in diverse land utilization patterns in North-East India: an implication for carbon sequestration. Environ Dev Sustain, 1–27. https://doi.org/10.1007/s10668-024-05037-y

Haldar S, Choudhury M, Choudhury S, Samanta P (2023) Trend analysis of long-term meteorological data of a growing metropolitan city in the era of global climate change. Total Environ Res Themes 7:100056. https://doi.org/10.1016/j.totert.2023.100056

Harley D, Mawson PR, Olds L, McFadden M, Hogg C (2018) The contribution of captive breeding in zoos to the conservation of Australia's threatened fauna. In: Recovering Australian threatened species: a book of hope, pp 281–294

Hilty JA, Keeley AT, Merenlender AM, Lidicker WZ Jr (2019) Corridor ecology: linking landscapes for biodiversity conservation and climate adaptation. Island Press

Hoffmann S (2022) Challenges and opportunities of area-based conservation in reaching biodiversity and sustainability goals. Biodivers Conserv 31:325–352

Hughes A, Auliya M, Altherr S, Scheffers B, Janssen J, Nijman V, Shepherd CR, D'Cruze N, Sy E, Edwards DP (2023) Determining the sustainability of legal wildlife trade. J Environ Manage 341:117987

Islam MZ (2023) A reality check of the global TX2 goals of doubling the Bengal Tiger (*Panthera tigris tigris*) population by 2022 in the Sundarbans Mangrove Forest. Contemp Probl Ecol 16:868–885

Jaya I, Satria F, Nugroho D, Sadiyah L, Buchary EA, White AT, Franklin EC, Courtney CA, Green G, Green SJ (2022) Are the working principles of fisheries management at work in Indonesia? Mar Policy 140:105047

Kang D (2022) A review of the habitat restoration of giant pandas from 2012 to 2021: research topics and advances. Sci Total Environ 852:158207

Kerry RG, Montalbo FJP, Das R, Patra S, Mahapatra GP, Maurya GK, Nayak V, Jena AB, Ukhurebor KE, Jena RC (2022) An overview of remote monitoring methods in biodiversity conservation. Environ Sci Pollut Res 29:80179–80221

Kremen C, Merenlender AM (2018) Landscapes that work for biodiversity and people. Science 362:eaau6020

Kumar D, Choudhury M, Rathore A (2020) Valuation of ecosystem services & benefits of Son Beel wetland in Assam, India: A case study of natural solutions to climate change & water: The Worldwide Online Climate Change Conference. Hamburg University, Germany

Kumar D, Choudhury M (2021). Recognizing economic values of wetland ecosystem services: a study of emerging role of monetary evaluation of Chandubi ecosystem and biodiversity. In: Wetlands conservation: current challenges and future strategies, pp 87–110. https://doi.org/10.1002/9781119692621.ch5

Lindenmayer DB, Lavery T, Scheele BC (2022) Why we need to invest in large-scale, long-term monitoring programs in landscape ecology and conservation biology. Curr Landsc Ecol Rep 7:137–146

Logue D, Grimes M (2022) Platforms for the people: enabling civic crowdfunding through the cultivation of institutional infrastructure. Strateg Manag J 43:663–693

Macdonald DW (2023) Biodiversity conservation: a very short introduction. Oxford University Press

Madliger CL, Franklin CE, Love OP, Cooke SJ (2020) Conservation physiology: applications for wildlife conservation and management. Oxford University Press, USA

Maiti PK, Maiti P (2023) Biodiversity: perception, peril and preservation. PHI Learning Pvt. Ltd.

Okorondu J, Umar NA, Ulor CO, Onwuagba CG, Diagi BE, Ajiere SI, Nwaogu C (2022) Anthropogenic activities as primary drivers of environmental pollution and loss of biodiversity: a review

Prakash T, Wijeratne A, Fernando P (2020) Human-elephant conflict in Sri Lanka: patterns and extent. Gajah 51:16–25

Raimi MO, Saliu AO, Babatunde A, Okon OG, Taiwo PA, Ahmed A-K, Loto O, Iyingiala A-A, Telu M (2022) The challenges and conservation strategies of biodiversity: the role of government and non-governmental organization for action and results on the ground. In: Biodiversity in Africa: potentials, threats and conservation. Springer, pp 473–504

Santiapillai C, Fernando P, Gunewardene M, A strategy for the conservation of the Asian elephant in Sri Lanka

Shrestha S, Sherpa AP, Lama S, Tamang M, Paudel PK (2021) Tigers at higher elevations outside their range: what does it mean for conservation? Eco Mont (j Prot Mountain Areas Res) 14:43–46

Shivanna K (2020) The sixth mass extinction crisis and its impact on biodiversity and human welfare. Resonance 25:93–109

Siddha S, Sahu P (2022) Impact of climate change on the river ecosystem. In: Ecological significance of river ecosystems. Elsevier, pp 79–104

Siddiqui JA, Bamisile BS, Khan MM, Islam W, Hafeez M, Bodlah I, Xu Y (2021) Impact of invasive ant species on native fauna across similar habitats under global environmental changes. Environ Sci Pollut Res 28:54362–54382

Singh J, Yadav P, Pal AK, Mishra V (2020) Water pollutants: origin and status. Sensors in water pollutants monitoring: role of material, pp 5–20

Van Wilgen BW, Strydom T, Simms C, Smit IP (2022) Research, monitoring, and reflection as a guide to the management of complex ecosystems: the case of fire in the Kruger National Park, South Africa. Conserv Sci Pract 4:e12658

Wei F, Fan H, Hu Y (2020) *Ailuropoda melanoleuca* (giant panda). Trends Genet 36:68–69

Yadav N, Upadhyay RK (2023) Global effect of climate change on seasonal cycles, vector population and rising challenges of communicable diseases: a review. J Atmos Sci Res

Ye Y (2017) Three essays on wildlife conservation programs: value, timing, and location

Young CN, Carlson JK (2020) The biology and conservation status of the oceanic whitetip shark (*Carcharhinus longimanus*) and future directions for recovery. Rev Fish Biol Fisheries 30:293–312

Chapter 6
Seeds of Sustainability: Promoting Food Security Through Eco-friendly Agriculture

Shivam Attri, Aanchal Saklani, Kamal Kant Joshi, Prernaa Sharma, Avnish Chauhan, Rekha Goswami, Bipin Sati, and Mariana Flores-García

Abstract Agriculture has been fundamental to the survival and growth of human civilization, transitioning societies from nomadic lifestyles to settled communities that evolved into cities, towns, and villages. The concept of food security is a critical global issue, impacting every individual. Historically, efforts to boost agricultural productivity have often led to environmental degradation, loss of biodiversity, and diminished ecosystem services, disproportionately affecting impoverished populations. In response, Eco-friendly agriculture farming that leverages a deep understanding of ecosystem services has emerged as a viable strategy for achieving global food security while curbing further environmental harm. The achievement of food security and sustainable livelihoods is a critical problem in emerging nations such as India, where smallholder farmers play a pivotal role in agricultural output. It is common to see organic farming portrayed as a means of enhancing smallholders' socioeconomic status, particularly in the Global South. Even though India has the largest population of organic farmers worldwide, there remains uncertainty about the real benefits of eco-friendly farming for smallholders in the country, especially regarding protecting livelihoods and guaranteeing food and nutrition security. Eco-friendly agriculture, encompassing practices that are less harmful to the environment compared to conventional farming, is widely viewed as a more sustainable approach. This chapter explores these dynamics, evaluating the role of sustainable and eco-friendly agriculture in promoting food security, challenges and opportunities in the Indian context.

S. Attri · A. Saklani · K. K. Joshi (✉) · P. Sharma · A. Chauhan · R. Goswami
Department of Environmental Sciences, Graphic Era Hill University, Dehradun, Uttarakhand, India
e-mail: kkjoshi@gehu.ac.inhoo.com

B. Sati
Department of Science and Technology, Govt. of Uttarakhand, Dehradun, Uttarakhand, India

M. Flores-García
Directora del Instituto Nacional de Arquitectura y Urbanismo, Universidad De La Salle Bajío, Leon, Mexico

M. Choudhury et al. (eds.), *Sustainable Synergy: Harnessing Ecosystems for Climate Resilience*, Climate Change Management,
https://doi.org/10.1007/978-3-031-77957-2_6

Keywords Organic farming · Pest control · Sustainability · Food security · Eco-friendly

Introduction

Food security is becoming more and more urban as a greater percentage of people on the planet reside in townships and cities. In the past few years it has been shown the cities and towns have likewise become the focal point of food safety, security and policy research, which formerly only included rural regions (Choithani et al. 2024). Reaching the sustainable development goal 2 (SDG2) of ending world hunger by 2030 is synonymous with achieving food security. Since the world became concerned about the oil and food crises in the 1970s, food security has been a major worldwide problem (McKeown 2006). Additionally, it was predicted that during the next 50 years, the issue of food security will continue to plague the world's population because rising food demand tends to conflict with falling food supply due to factors like economic growth, the need for energy crops, change in climate, water scarcity, and environment degradation (Rosegrant and Cline 2003). Increasing agricultural productivity is still the primary means of achieving food security. Agricultural innovation, on the other hand, is a powerful tool for raising agricultural productivity, particularly over the long term. Technological advancements in agriculture open the door to high-yield crops, reduced crop losses, and more effective use of scarce resources, and the ultimate shift from subsistence farming with low yield to high productivity. Historically, attempts to improve the yields and production have had a negative impact on the soil biodiversity and degrade the environment. Sustainable agriculture practices and including the knowledge of ecosystem services, is not only an effective solution to ensuring the food security for everyone even preventing the additional environment deterioration (Rehman et al. 2022). One of the agricultural industries with the quickest growth is Eco-friendly farming. Its cornerstone is the idea that healthy soil supports healthy plants, animals, and people. A globally acknowledged production method, organic agriculture is founded on the values of justice, fairness, health, and environment. It is propelled by customer demand and governed by manufacturing standards that are legislated in many nations worldwide. With small- and large-scale farmers, vast and intense organic production methods, technologically sophisticated and limited production, and a wide range of growing regions from tropical to desert to sub-arctic, organic farming is genuinely global in character. In order to help us advance toward sustainable food systems, organic agriculture provides a distinctive model with its guiding principles and standards (Hammermeister 2020).

This chapter explores how environmentally friendly agriculture might help ensure the world's food supply while also preserving healthy ecosystems that provide humanity with other important benefits. The idea that agricultural production systems are distinct from other natural ecosystems is too prevalent, and the ways in which services might enter and exit agricultural production systems from surrounding ecosystems have received little consideration. Along with outlining the trade-offs and synergies between ecosystem services, this chapter also focuses on food and nutrition security implications of urbanization processes and also identifies existing knowledge gaps and suggests topics for further investigation.

In the global context of modern era various types of biowaste of these food waste is one of the popular. Generally, the Food waste is a composition of complex carbohydrates, proteins, fatty lipids, organic acids, enzymes, and nutraceuticals and it can be defined as any edible or inedible food loss from the food-supply chain (Dey et al. 2024). The storages of food wastes have became a major environmental and economic problem, with reference to environment food waste release greenhouse gases (GHG). It is estimated total 20% of the total global production of GHG, including CH_4, N_2O, and CO_2 and also responsible reduced the ground water quality and decrease the land-use area.

Valorization of food waste is an expanding industry due to increasing public awareness of indiscriminate disposal of food waste and its environmental impacts. A circular bioeconomy has been adopted by many policymakers, and their decisions have led to sustainable initiatives within the social, commercial, and industrial sectors of society However, these decisions are subjected to scrutiny, because valorisation is still in its early stages (Mak et al. 2020; Chowdhury et al. 2024).

Generally, the fertilizers are applied to increase plant growth and crop yield in the agricultural fields. The Food waste fertilizers, such as composts, digestates, and dehydrates are helpful to increase soil fertility and microorganisms. Especially in European countries the utilization of anaerobic digestion of animal manure and sewage sludge at full-scale is relatively common while anaerobic digestion of food waste is not common. Thus, public awareness and strong policies regarding the food wastes valorization is must be introduced in all the nations to sustain the environment, agriculture, and food security.

Key Principles of Organic Agriculture

Humans depend on soil to supply essential functions such as agricultural production maintenance and assurance, habitat provision for creatures, and environmental health (Creamer et al. 2022). Soil health is facing previously unheard-of difficulties as a result of the increase of agricultural management techniques intended to maintain global food security. This challenge raises questions about the roles and services of soil, not just for food security but also for the increasing demands brought on by dwindling global resources, severe environmental problems, degraded ecosystems, and climate change (Qiao et al. 2022). The various management strategies affect on the soil health, and incorporate soil microorganisms, even before noticeable changes in physical or chemical markers, because they can react quickly and early to environmental changes and pressures (Lehmann et al. 2020). Since improper land management has blocked drainage channels or resulted in farmers' neglecting their responsibilities. To preserve the soil fertility is crop diversity and adds rice straws and animal manure into the soil to increase organic matter in addition continue use of vermicompost, guano, bio-char, farm-yard manure, green manure, and organic liquid manure in order to promote sustainable agriculture (Li et al. 2023; Paul et al. 2019).

Sustainable agriculture emphasizes diversification, integration, and knowledge-based management to enhance the well-being of the entire system, including people and the environment. This sustainable agriculture practices focus on adapting to the environmental limits and creating resilient, self-sustaining communities, rather than just increasing a production.

The problem of food security necessitates the capacity to cope with growing food shortages for a global population that is constantly growing. With an estimated 1.7 billion more people expected to live on Earth by the year 2050, humankind is putting increasing strain on the limited resources that are used to grow our food (Mc Carthy et al. 2018). A multifaceted approach to agricultural sustainability is required to prevent harm to social, economic, and environmental integrity from the demand for increased food production to feed the world's rapidly expanding population. Climate change is a serious threat to the current state of global food security as it will negatively affect agricultural yields, especially in low-income countries. Increased agricultural intensification to produce more food from existing cropland has put environmental sustainability at risk due to increased emissions, loss of biodiversity, and soil health from increased use of chemical fertilizers and pesticides. The importance of social, economic, and environmental sustainability to agricultural sustainability and food security is discussed in this chapter. It goes on to discuss the state of food security today and suggests ways that sustainable intensification can lessen the negative effects that intensification has on the agricultural industry's social, environmental, and financial aspects. Considering the changing climate, it also emphasizes mitigating techniques for achieving food security and safety (Farooq et al. 2019).

Ultimately, sustainable agriculture calls for a systems-based approach to address modern environmental, economic, and social challenges, moving beyond the component-focused methods of the conventional and industrial model. Figure 6.1 shows the basic principles for soil health.

Principles of Eco-friendly Agriculture

One of the most difficult problems in the pursuit of sustainable development is the unchecked use of pesticides, including fungicides, herbicides, and insecticides. Pesticides that are heavily utilized pose a risk to human health, the environment, and food safety. Plants and soils are treated directly with pesticides. Inadvertent pesticide releases have the potential to cause long-term pesticide persistence in the environment. Air, water, and soil pollution may be avoided with effective pesticide management (Ozkara et al. 2016). The natural method of controlling pests through the interplay of predators, parasitoids, and pests to maintain the balance of each

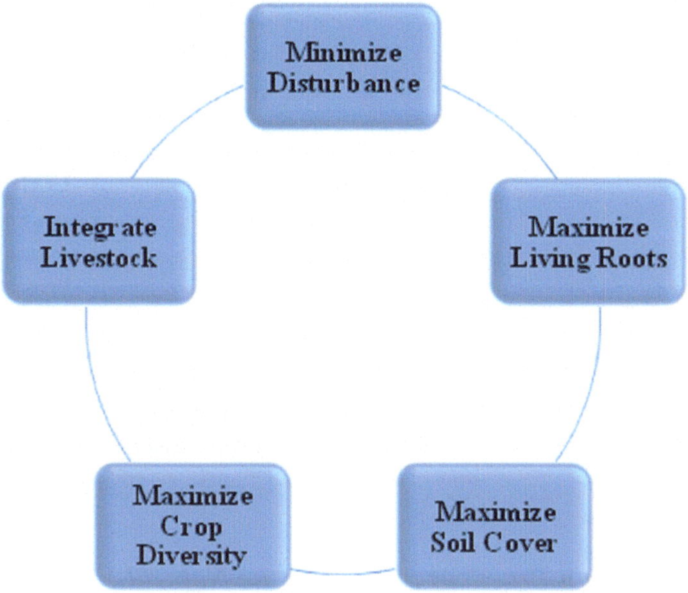

Fig. 6.1 Basic principles for soil health

population has been disturbed by excessive and improper use of insecticides (El-Wakeil et al., 2013).

In last fifty decades the use of pesticide has been increased about 0.27% in India while the maximum 0.19% was calculated in the year of 1994–1995. During green revolution '1968–78' to increase the food production and food safety maximum pesticide were used in India. Now keep the effects of pesticide further, agriculture sustainability promoting the organic pesticide.

The various pest and disease are one of the major effective causes of the crop production and modern agriculture practices (Fig. 6.2). Using the proper management techniques it can be minimized and reduced. The following categories for simplicity of comprehension and efficient application in the control of insect pests and illnesses under organic farming:

Modification of Cultural Practices

Among the earliest methods of controlling pests are cultural practices. Many modern preventative methods employed in conventional and organic farming, such as crop rotation, soil health management, and the use of insect-resistant plants, have their origins in traditional agriculture (Haldhar et al. 2017). Figure 6.2 shows crop rotation in the agriculture field.

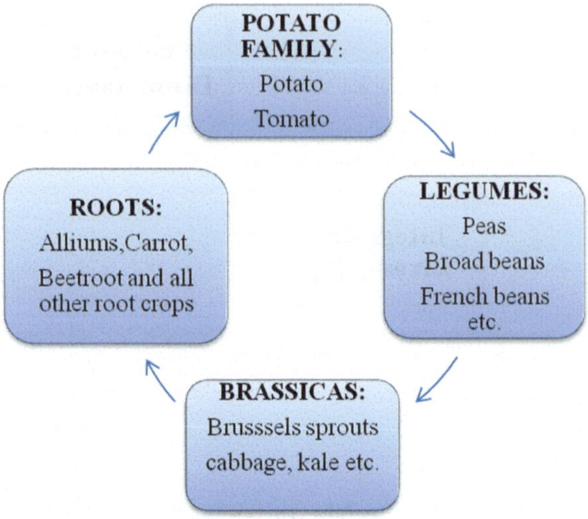

Fig. 6.2 Crop rotation in the agricultural area

The Conservation Practices to Restore the Natural Enemies

Conservation of natural enemies is modifying the environment to improve the longevity, fertility, and behaviour of natural enemies in order to boost their efficacy. Examples of such modifications include the installation of hedge rows and shelter belts (Haldhar et al. 2017).

Use of Biological Control Agents

When it comes to controlling insect pests in an environment devoid of insecticides, methods such as inundative and inoculative release or the application of biological control agents including parasitoids, insect diseases, and insect predators will be more effective. If an unexpected epidemic of insects occurs, these chemicals can be employed as curative control measures (Haldhar et al. 2017). Table 6.1 shows bio control agents and their effectiveness in crops.

Use of Botanicals and Their Mixtures

When all other approaches of controlling insect pests have failed, the use of botanicals and other pesticides with a mineral origin is reserved for last resorts in organic agriculture. However, in India to promote the organic farming the government of India has been started the ***National Programme for Organic Production*** (NPOP). The major objective of this programme is to strict regulation of chemicals which are

Table 6.1 Bio-control agents and their effectiveness in crops

S. No.	Biocontrol agents	Species name	Effective against	Crops
1	Bacteria	*Bacillus thuringiensis*	Lepidopteran pests	Cotton, sunflower and vegetables
2	Entomopathogenic fungi	*Metarhizium anisopliae*, *Beauveria bassiana* and *Verticilium lecanii*	Coleopteran grubs Lepidopteran and Coleopteran Aphids and whiteflies	Coconut, Cotton and green house vegetables
3	Insect predators	Lady bird beetles and *Chrysoperia* spp.	Aphids, Whiteflies and mealy bugs	Fruits and vegetables
4	Insects' parasitoids	*Trichogramma* spp. and *Chelonis blackburni*	Lepidopterans	Sugarcane and tomato
5	Entomopathogenic nematodes	*Heterorhabditis bacteriophora* and *Steinernema carpocapsae*	Coleopteran and Lepidopterans	Sugarcane and plantation crops
6	Entomopathogenic viruses	Nuclear polyhedrosis virus (NPV) and Granulosis virus	Helicoverpa and other lepidopterans	Pulses and vegetables

using in pest management in organic farming and registration of chemicals using in organic crop cultivation. In this regards, the ***Panchagavya*** and ***Dasagavya*** mixtures are derived from animal products for organic pest control (Haldhar et al. 2017).

Use of Pheromones and Other Attractant

In a vast and expansive universe, insects are little beings. They have developed several methods for locating one another for mating. Some insects have eye-catching hues; others produce sounds as loud as a chainsaw. Many insects use chemical "pheromones," to lure other insects for matting. This allows the insects to communicate across great distances.

Pests can be monitored, mating can be disrupted, vast numbers of adults can be trapped, an insect infection may be dispersed, or pests can be lured to eat poisoned bait using pheromones and other chemical attractants (Haldhar et al. 2017).

Use of Organic Pesticides

Applying biological or botanical material or a substance on the national list of synthetic substances approved for use in organic crop production can help prevent, suppress, or control pest populations when nonchemical methods outlined in the organic system plan are insufficient to keep them from growing to an economically damaging level. For instance, the fermentation product of the soil-dwelling actinomycete *Saccharopolyspora spinosa* is the pesticide Spinosad (Haldhar et al. 2017).

Impacts

Economic and Social Impacts

The International Fund for Agriculture Development (IFAD) reports shows that India's organic production has been increasing gradually. In India, almost 2.5 million hectares of land are used for organic farming. Among the 187 nations that practice organic agriculture.

With 2.30 million acres, India accounts for 30% of global organic growers. 27 total acres under organic cultivation, 59,660 total farmers (1,160,650 PGS and 1,599,010 India Organic), 1703 total processors, and 745 dealers comprise the total crop (**NCONF**; National Centre for Organic and Natural Farming). Growing consumer demand for sustainably produced food and worries about the environmental effects of chemical-intensive farming have led to the emergence of organic farming as a substantial alternative to conventional agriculture.

- **Premium Costs**: Comparing organic to conventional items, they frequently command higher pricing on the market. With organic farming, labor, organic seed, and natural fertilizers are often more expensive. This premium can help offset some of these expenses.
- **Reduced Input Expenses**: In terms of synthetic chemicals, herbicides, and fertilizers, organic farming usually has cheaper input prices, even when labor expenses may be higher. Natural substitutes are frequently used by farmers and can ultimately be less expensive.
- **Aid and Subsidies from the Government**: Governments provide grants, subsidies, and other financial incentives to promote organic farming in several nations. These can lessen a portion of the initial investment and transition costs associated with organic farming to the farmers.
- **Optimal Soil Health for Extended Productivity**: Long-term improved yields can result from organic agricultural techniques, which prioritize preserving the health of the soil. Better soils reduce reliance on outside resources, which lowers costs and increases production over the long term

- **Diversification Opportunities**: Crop diversity and rotation are common practices in organic farming, which can generate several revenue streams. This can raise agricultural profitability overall and lower the chance of a crop failing completely. Beyond its effects on the ecology and economy, organic farming contributes significantly to social benefits and increased community involvement.

Sustainable Farming Techniques

Crop Diversification and Rotation

The crop diversification has a potential to mitigate the adverse environmental effects and the loss of biodiversity while maintaining cropping systems' production. However, there is still a lack of consensus on the topic and methods for doing more thorough study. In contrast to specialized farming, crop diversification refers to an effort to enhance crop variety using techniques like crop rotation, multiple cropping, or intercropping in order to boost production, stability, and the provision of ecosystem services (Garbach et al. 2017).

The crop diversification and rotation promote to increase the genetic, temporal, and spatial diversity of cropping systems. Temporal shifts and ranges of phenological stages (important for biodiversity and climate change adaptation), more frequent or continuous soil cover, and a greater variety of management techniques, decrease in labor peaks and economic risk, are all effects of diversification and also, compared to simplified systems, this form of cropping system produced yields that were equal to or higher, with less of an impact on the environment (Davis et al. 2012).

Water Management

Worldwide, water is regarded as the most important resource for the development of sustainable agriculture. In the upcoming years, there will be a rise in irrigated areas as freshwater resources are diverted from agriculture to fulfill the growing demands of industry and home consumption. Moreover, the effectiveness of irrigation is extremely low, as the crops utilize less than 65% of the water that is applied. Understand scarcity conditions and climate change considerable effort has been devoted over time to introduce policies aiming to increase water efficiency based on the assertion that more can be achieved with less water through better management. Sustainable water management in agriculture and environmental protection are linked to agricultural operations like managing soil, applying fertilizer, watering crops, and controlling pests and diseases. Adopting sustainable water management in the Mediterranean is not just a technological issue therefore numerous other factors to consider, such as rural communities' social behaviors, financial limitations, and legal framework, which may encourage the adoption of some measures but not others.

Soil Conservation Techniques

Soil degradation is one of the most pressing environmental concerns. Agricultural activity and its intensification are closely linked to a number of processes that cause soil degradation, like erosion, compaction, salinization, pollution, and loss of both organic matter and soil biodiversity (Gogoi et al. 2024). Scientists and farmers alike face a challenge in the development and implementation of sustainable agronomic techniques that can maintain and improve the physical, chemical, and biological characteristics of soils and enhance the functioning of agro ecosystems (Castellini et al. 2021). In Addition, agroforestry practices, mulching, and strip cropping are helpful to control soil erosion and improve organic matter in soil. As well, utilization of biochar, integrated pest management, organic amendments, and integrated nutrient management also help to improve the soil structure and nutrient content.

Integrated Pest Management (IPM)

Integrated Pest Management (IPM) is a critical approach within sustainable agriculture practices that seeks to minimize the reliance on chemical pesticides while effectively manages the pest populations. It is believed that technological developments in pest control will yield chemical control agents as compared to conventional pesticides. This practice is more effective less harmful, less persistent, and less likely to seep into the soil. Furthermore, farmers will be able to apply pesticides more effectively and selectively, as well as the expanding use of monitoring and knowledge-based systems. IPM provides understanding about the limitation of pesticides (i.e. when, where, and how much need) chemical. It also provides a well-rounded approach that safeguards crops, conserves beneficial creatures, and upholds the health of ecosystems (Viatte 2001).

Case Study: Community-Supported Agriculture (CSA)

The concept of Community-Supported Agriculture (CSA) serves as an example of how organic farming may improve relationships within the community. A CSA gives farmers upfront financing and a guaranteed market for their food by allowing customers to purchase portions of the farm's crop in advance. Members receive seasonal, fresh, organic fruit on a regular basis in exchange.

1. **Building Community Relationships**—Stronger social relationships are formed between farmers and customers as a result of organic agricultural techniques, which frequently include direct contact with local communities
2. **Enhancing Physical and Mental Welfare**—Producing food free of artificial chemicals, pesticides, and genetically modified organisms (GMOs) is one way that organic farming improves community health and wellbeing.

3. **Encouragement of Rural Development and Economic Rejuvenation**—Supporting rural development and economic regeneration may be greatly aided by organic farming. Organic farming has the potential to boost rural communities that have been adversely affected economically by the industrialization of agriculture by bringing in new enterprises, producing cash, and creating employment.
4. **Promoting Knowledge Sharing and Cultural Preservation**—Traditional agricultural methods and expertise are frequently used and preserved in organic farming. This might involve using heritage seeds, local agricultural practices, and conventional approaches to pest and soil management. Organic farming guarantees that important agricultural knowledge is passed down through the generations and contributes to the preservation of cultural heritage by upholding these practices.
5. **Technological Innovations in Eco-friendly agriculture**—Traditional agricultural methods and expertise are frequently used and preserved in organic farming. This might involve using heritage seeds, local agricultural practices, and conventional approaches to pest and soil management. Organic farming guarantees that important agricultural knowledge is passed down through the generations and contributes to the preservation of cultural heritage by upholding these practices. An ecologically beneficial approach to regulating agricultural production inputs can be achieved through precision agriculture. It is able to target fertilizer, seed, and chemical rates for soil and other variables by applying site-specific knowledge. Physical inputs are replaced with knowledge and information through precision agriculture. Consistent with the intuitive notion that precision agriculture should lessen environmental loading by applying fertilizers, pesticides, and precision agriculture can support long-term sustainability of production agriculture in a number of ways. The environmental benefits of precision agriculture stem from the more focused application of inputs, which minimizes losses from overuse as well as losses from weed escapes, insect damage, and nutrient imbalances. A decrease in the emergence of pesticide resistance is one of the additional advantages (Bongiovanni and Lowenberg-DeBoer 2004). Utilizing remotely sensed imagery from satellites, aerial platforms, and GPS-tagged drones, geospatial technology plays a major part in improving agricultural practices, crop development, and yields. Geospatial technology allows us to map and track crop and soil conditions as well as efficiently manage, safeguard, and conserve biodiversity worldwide. In order to maximize yields and agricultural productivity, remote sensing applications take into account ecological and environmental characteristics, soil variables, crop conditions, and plant-soil variety (Pandey and Pandey 2023).Sustainable agriculture aims to protect natural resources while securing the availability of food and energy in the future. In order to support energy demands and reduce environmental damage associated with agriculture, renewable energy is essential to sustainable agriculture. It reduces reliance on fossil fuels and generates electricity from agricultural waste (Bathaei and Štreimikienė, 2023; Singh et al. 2022).

Policy and Economic Support for Sustainable Agriculture

Government Policies and Incentives

Sustainable agriculture is greatly aided by government regulations and financial incentives, which provide farmers the resources and assistance they need to implement eco-friendly techniques. Financial incentives for cutting back on chemical inputs like synthetic fertilizers and pesticides, as well as subsidies for sustainable agricultural practices like organic farming, agro-forestry, and conservation tillage, can be included in effective policy frameworks. In order to promote the use of sustainable technology like precision agricultural instruments, renewable energy sources, and effective irrigation systems, governments can also enact tax incentives, grants, and low-interest loans. Support for education and extension programs, together with research and development funding, can also aid in knowledge dissemination and capacity building among farmers. Governments may safeguard natural resources and encourage sustainable land management practices by enacting explicit restrictions.

Economic Viability and Market Access

All agricultural enterprises must prioritize economic viability, which has grown in importance and popularity as a study topic. The EU's 2014–2020 agriculture strategy, which prioritizes increasing the small- and mid-sized farms' economic viability, provides the basis for the topic's importance. Rather than growing farms larger, the goal of increasing farm economic viability is to encourage entrepreneurship among farms, provide new employment opportunities, and take action against a decline in rural population. Smallholders' capacity to participate in the market is becoming more and more important to their ability to increase their incomes, yet they face several obstacles due to market failures in developing countries' rural markets. It is especially crucial to comprehend how group efforts might assist in resolving inefficiencies, issues with coordination, or obstacles to market access (Markelova et al. 2009).

Education and Training

The advancement of sustainable agriculture depends heavily on education and training since it provides communities, agricultural professionals, and farmers with the information and abilities they need to adopt ecologically friendly techniques. Managing soil health, integrated pest control, crop diversification, water conservation strategies, and the application of organic inputs are just a few of the subjects that good education programs ought to address. There are several ways to deliver training that will be accessible to a wide range of people, including workshops, internet platforms, farmer-to-farmer exchanges, extension services, and field demonstrations. Agronomists and farmers of the future can be better prepared with cutting-edge techniques and a thorough grasp of ecological concepts if sustainability is integrated into official agricultural education, such as university and vocational programs. Furthermore, chances for ongoing professional development for educators and extension agents can aid in bridging the knowledge gap between scientific research and practical application. Initiatives in education and training strengthen farmers' ability to implement sustainable practices, increase productivity, strengthen the resilience of agricultural systems, and promote long-term food security.

Challenges and Opportunities

As far as we are aware, certification of "process" is utilized in organic agricultural standards instead of "outcomes". Organic agriculture is recognized for having distinct characteristics and unique outcomes compared to other farming systems. These differences are important to motivate the cultivators to use organic farming methods, despite these challenges they may face, such as limited input options, pest control difficulties, sustaining the yields, market competition, certification expenses, and stringent record-keeping requirements. The consumers of organic products are willing to pay higher prices, expecting that these products or their production processes offer unique benefits compared to conventional options.

The most frequent metric used to assess the effectiveness of agricultural systems is yield. The extent of the difference depends on the crop, the growing environment, if best practices are being compared, and whether the comparison is being conducted in rich or developing nations. Yields in organic cultivation can sometimes be lower than non-organic yields (Seufert et al. 2012).

Conclusion

Eco-friendly agriculture is a globally recognized system that brings together diverse farmers under common standards, making it a valuable model for promoting food security through eco-friendly practices. Organic farming not only establishes a distinct identity for consumers but also helps retailers to understand market demands. As a sustainable food system, organic agriculture offers a platform for evaluating the balance between achieving high yields and minimizing environmental impacts.

An integrated strategy that incorporates market access, economic viability, government backing, thorough education, and training is crucial for sustainable agriculture. Farmers may embrace sustainable technology and eco-friendly practices with the help of government rules and financial incentives like grants, tax breaks, and subsidies. It is critical to maintain small- and mid-sized farms' financial stability since doing so improves their access to markets, encourages entrepreneurship, and boosts employment in rural areas. Additionally, education and training initiatives equip farmers with the know-how to successfully adopt sustainable methods, bridging the gap between research findings and practical application. When combined, these tactics provide a strong foundation for developing resilience, improving production, and supporting sustainable agriculture.

References

Bathaei A, Štreimikienė D (2023) Renewable energy and sustainable agriculture: review of indicators. Sustainability 15(19):14307. https://doi.org/10.3390/su151914307

Bongiovanni R, Lowenberg-DeBoer J (2004) Precision agriculture and sustainability. Precis Agric 5:359–387. https://doi.org/10.1023/B:PRAG.0000040806.39604.aa

Castellini M, Diacono M, Gattullo CE, Stellacci AM (2021) Sustainable agriculture and soil conservation. Appl Sci 11(9):4146. https://doi.org/10.1371/journal.pone.0047149

Choithani C, Jaleel CP, Rajan SI (2024) Rural-urban transition and food security in India. Glob Food Secur 42:100780

Chowdhury R, Barman S, Choudhury M, Kim KH, Bhattacharya SS (2024) Earthworm modifies microbial community and functional genes for lignocellulosic waste valorization: isolating plant-growth-promoting bacteria via next generation sequencing. Int Biodeterior Biodegradation 193:105854. https://doi.org/10.1016/j.ibiod.2024.105854

Creamer RE, Barel JM, Bongiorno G, Zwetsloot MJ (2022) The life of soils: integrating the who and how of multi functionality. Soil Biol Biochem 166:108561

Davis AS, Hill JD, Chase CA, Johanns AM, Liebman M (2012) Increasing cropping system diversity balances productivity, profitability and environmental health. PLoS ONE 7(10):e47149. https://doi.org/10.1371/journal.pone.0047149

Dey S, Santra M, Choudhury M, Ghosh AR, Samanta P (2024) Food waste generation and its industrial utilization: an overview. Environ Sci Pollut Res, 1–20. https://doi.org/10.1007/s11356-024-34252-3

El-Wakeil N, Gaafar N, Sallam A, Volkmar C (2013) Insecticides development of safer and more effective technologies In: Stanislav T (ed) Side effects of on natural enemies and possibility of their integration in plant protection strategies. Intech Open. https://doi.org/10.5772/54199

Farooq M, Rehman A, Pisante M (2019) Sustainable agriculture and food security. Innovations Sustain Agric 3–24

Garbach K, Milder JC, DeClerck FAJ, Montenegro de Wit M, Driscoll L, Gemmill-Herren B (2017) Examining multi-functionality for crop yield and ecosystem services in five systems of agroecological intensification. Int J Agric Sustain 15(1):11–28. https://doi.org/10.1080/14735903.2016.1174810

Gogoi N, Sarma A, Choudhury M, Samanta P, Faizan M, Sadhak S (2024) Soil heavy metal pollution and ecological risk assessment in disturbed and undisturbed soil of Morigaon, Assam. Discover Environ 2(1):13. https://doi.org/10.1007/s44274-024-00039-z

Haldhar SM, Jat GC, Deshwal HL, Gora JS, Singh D (2017) Towards organic agriculture. In: Gangwar B (ed) Insect pest and disease management in organic farming. New Delhi, India

Hammermeister AM (2020) Introduction to agroecology. In: Organic agriculture: a model for sustainability, pp 279–299. https://doi.org/10.1007/978-981-15-8836-5_20

Lehmann J, Bossio DA, Kogel-Knabner I et al (2020) The concept and future prospects of soil health. Nat Rev Earth Environ 1:544–553. https://doi.org/10.1038/s43017-020-0080-8

Li X, Qiao L, Huang Y, Li D, Xu M, Ge T, Meersmans J, Zhang W (2023) Manuring improves soil health by sustaining multifunction at relatively high levels in subtropical area. Agric Ecosyst Environ 353:108539. https://doi.org/10.1016/j.agee.2023.108539

Mak TM, Xiong X, Tsang DC, Iris KM, Poon CS (2020) Sustainable food waste management towards circular bioeconomy: policy review, limitations and opportunities. Bioresour Technol 297:122497. https://doi.org/10.1016/j.biortech.2019.122497

Markelova H, Meinzen-Dick R, Hellin J, Dohrn S (2009) Collective action for smallholder market access. Food Policy 34(1):1–7. https://doi.org/10.1016/j.foodpol.2008.10.001

Mc Carthy U, Uysal I, Badia-Melis R, Mercier S, O'Donnell C, Ktenioudaki A (2018) Global food security—issues, challenges and technological solutions. Trends Food Sci Tech 77:11–20. https://doi.org/10.1016/j.tifs.2018.05.002

McKeown D (2006) Food security: implications for the early years. Background paper. Toronto Public Health

Ozkara A, Akyıl D, Konuk M (2016) Pesticides, environmental pollution, and health. In: Larramendy L (ed) Environmental health risk-hazardous factors to living species. Intech Open. https://doi.org/10.5772/63094

Pandey PC, Pandey M (2023) Highlighting the role of agriculture and geospatial technology in food security and sustainable development goals. Sustain Dev 31(5):3175–3195. https://doi.org/10.1002/sd.2600

Paul S, Choudhury M, Deb U, Pegu R, Das S, Bhattacharya SS (2019) Assessing the ecological impacts of ageing on hazard potential of solid waste landfills: a green approach through vermitechnology. J Clean Prod 236:117643. https://doi.org/10.1016/j.jclepro.2019.117643

Qiao D, Li N, Cao L, Zhang D, Zheng Y, Xu T (2022) How agricultural extension services improve farmers' organic fertilizer use in China? The perspective of neighborhood effect and ecological cognition. Sustainability 14(12):7166. https://doi.org/10.3390/su14127166

Rehman A, Farooq M, Lee DJ, Siddique KH (2022) Sustainable agricultural practices for food security and ecosystem services. Environ Sci Pollut Res 29(56):84076–84095. https://doi.org/10.1007/s11356-022-23635-z

Rosegrant MW, Cline SA (2003) Global food security: challenges and policies. Science 302(5652):1917–1919. https://doi.org/10.1126/science.1092958

Seufert V, Ramankutty N, Foley JA (2012) Comparing the yields of organic and conventional agriculture. Nature 485(7397):229–232. https://doi.org/10.1038/nature11069

Singh S, Singh P, Sharma A, Choudhury M (eds) (2022) Agriculture waste management and bioresource: the circular economy perspective. Wiley. https://onlinelibrary.wiley.com/doi/book/10.1002/9781119808428

Viatte G (2001) Adopting technologies for sustainable farming systems: an OECD perspective. In: Adoption of technologies for sustainable farming systems Wageningen workshop proceedings, vol 14. OECD, Paris, France, pp 14–23

Chapter 7
Ecosystem Resilience and Water Resources: A Synergistic Approach

Prasann Kumar, Joginder Singh, and Anmoldeep Kaur

Abstract This chapter urgently delves into the intricate relationship between ecosystem resilience and water resources, a topic of immediate importance in the face of climate change. It underscores the vital role of ecosystems in upholding water availability and quality. Ecosystems function as natural infrastructure by regulating hydrological cycles, mitigating floods and droughts, and purifying water. This enhances the resilience of both natural and human communities. Understanding and utilising these ecosystem-water connections are pivotal for efficient water resource management and climate adaptation strategies. This chapter draws on insights from various fields to summarise existing knowledge and inspire new methods to leverage ecosystem resilience for sustainable water resource management in a changing climate, emphasising the immediate need for attention and action in this crucial area.

Keywords Ecosystem resilience · Water resource management · Hydrological cycle · Ecology

Introduction

Ecological resilience in water resources refers to the ability of natural systems to withstand and recover from disruptions while maintaining their fundamental roles and provisions for water accessibility, quality, and governance. This concept means

P. Kumar (✉)
Department of Agronomy, School of Agriculture, Lovely Professional University, Phagwara, Punjab 144411, India
e-mail: prasann0659@gmail.com

J. Singh
Department of Botany, Nagaland University, Nagaland, India

A. Kaur
Department of Agriculture, Royal Agricultural University, Cirencester GL7 6JS, Gloucestershire, UK

M. Choudhury et al. (eds.), *Sustainable Synergy: Harnessing Ecosystems for Climate Resilience*, Climate Change Management,
https://doi.org/10.1007/978-3-031-77957-2_7

ecological systems can adapt to climate change and human influences like pollution and land use. The resilience of ecosystems in water resources is crucial to maintaining the hydrological cycle and providing water-related services that humans and ecosystems need. For instance, wetlands, rivers, lakes, and aquifers significantly regulate water flow quantity, timing, and characteristics (Choudhury et al. 2021). Wetlands, in particular, absorb and retain water during heavy rain and release it slowly during dry periods, which helps reduce downstream flooding and recharge groundwater. Riparian zones near rivers and streams provide a variety of flora and fauna habitats and act as natural filtration systems to remove pollutants and improve water quality (Kumar et al. 2020). On the other hand, forests protect watersheds, control runoff, reduce soil erosion, and maintain stream flows. They also replenish groundwater and strengthen water resources with their complex root systems. Forests also absorb carbon, reducing climate change and stabilizing hydrology. Ecosystems regulate water quality through sediment, nutrient, and contaminant filtration from surface runoff and groundwater, improving water purity and suitability for drinking, agriculture, and industry. Coastal mangrove forests, coral reefs, and seagrass beds also play a crucial role in trapping pollutants, mitigating storm surges, and facilitating fisheries, ensuring food security and livelihoods. The resilience of ecosystems is therefore crucial to sustainable water resource management because healthy and undamaged ecosystems ensure water supply durability, dependability, and excellence. Recognizing and protecting ecosystems' functional integrity is essential to the sustainability of water systems and their benefits to the environment and society. Rivers, lakes, wetlands, and aquifers help maintain water availability and quality, promote biodiversity, and provide vital services to human communities. Ecosystems regulate the hydrological cycle, store and provide fresh water, and filter contaminants, ensuring clean, reliable water for various uses. Water ecosystems control water flow amount and timing, reducing floods, droughts, and other extreme hydrological events (Kumar and Choudhury 2021). Wetlands absorb excess water during heavy rain and release it slowly during dry periods. This process reduces downstream flooding and recharges groundwater. Rivers and lakes retain water during high precipitation and gradually release it during low precipitation, maintaining stream flows and supporting ecological systems and human livelihoods that depend on water (Choudhury et al. 2022). Water ecosystems purify and filter water by removing sediments, nutrients, and pollutants from surface runoff and groundwater. Wetlands, riparian zones, and mangrove forests filter pollutants and improve water quality before it reaches downstream consumers. These ecological systems protect drinking water, fisheries, aquaculture, and biodiversity by maintaining healthy aquatic habitats. Water ecosystems support human welfare and economic growth by providing agricultural irrigation, hydropower, navigation, and recreation. Food production and livelihoods for a large global population depend on rivers and reservoirs for irrigation. Hydropower plants generate renewable energy and reduce greenhouse gas emissions by using water flow kinetic energy. Tourism, fishing, boating, and other outdoor activities are possible in aquatic environments. Coral reefs, mangroves, and seagrass beds attract tourists and support coastal communities by providing habitats for various marine species, protecting shorelines from erosion, and promoting recreational fisheries and diving.

Fig. 7.1 Ecosystem resilience and water resources. *Source* Created by authors

Water ecosystems provide vital services to humans, maintain water availability and quality, and encourage biodiversity (Sharma and Choudhury 2021). Accepting the importance of water resources and implementing sustainable management strategies are essential to their resilience and sustainability and the interconnected ecological systems that depend on them, empowering the audience to take action and implement these strategies (Fig. 7.1). Table 7.1 shows ecosystem resilience and water resources.

Ecosystem Services for Water Resources

Hydrological cycles are influenced by forests, wetlands, and rivers, which control water flow in watersheds and landscapes. Forests, such as marshes, swamps, and bogs, regulate water flow by absorbing and releasing water, reducing floods and droughts, and preserving river baseflow. Conversely, wetlands regulate hydrological cycles by absorbing and storing excess water during intense rainfall and gradually releasing it during low precipitation. They also help recharge groundwater by infiltrating excess water into the soil, improving groundwater storage and accessibility. Wetlands' dense vegetation and microbial communities filter surface runoff and groundwater, improving water quality and aquatic environments. Rivers transport water, sediment, and nutrients, regulate water flow, maintain aquatic environments, and provide

Table 7.1 Ecosystem resilience and water resources

Aspect	Description
1. Definition	Definition of ecosystem resilience and its relevance to water resources
2. Importance	Importance of prioritising ecosystem resilience in climate adaptation strategies
3. Future directions	Future directions for research and policy regarding ecosystem resilience and water resources
4. Ecosystem services	Regulation of hydrological cycles by forests, wetlands, and rivers
5. Water purification	Processes in natural ecosystems that contribute to water purification
6. Role of biodiversity	Biodiversity's role in maintaining ecosystem resilience and water quality
7. Climate change impacts	Shifts in precipitation patterns and hydrological extremes due to climate change
8. Water scarcity	Implications of climate change on water scarcity and water quality
9. Ecosystem management	Ecosystem-based approaches to water resource management
10. Integration	Integration of ecosystem services into water infrastructure planning
11. Ecosystem-based initiatives	Initiatives focused on managing water resources based on ecosystem principles
12. Policy frameworks	Policy frameworks for integrating ecosystem resilience into water management strategies
13. Stakeholder roles	Role of stakeholders in promoting ecosystem-based solutions for water resilience
14. Emerging technologies	They are emerging technologies and methodologies for enhancing ecosystem-water resilience
15. Research gaps	Gaps and priorities for advancing understanding and practice in ecosystem-based water management
16. Remote Sensing	Use of remote sensing for monitoring ecosystem health and water resources
17. Geographic information systems (GIS)	Role of GIS in spatial analysis, mapping, and decision support for ecosystem-water interactions
18. Artificial intelligence (AI)	Application of AI techniques in modelling and predicting ecosystem responses to environmental change
19. Sensor technologies	Use of sensor technologies for real-time monitoring of water quality and quantity in natural ecosystems

(continued)

Table 7.1 (continued)

Aspect	Description
20. Ecohydrological modeling	Modeling approaches that integrate ecological and hydrological processes to simulate ecosystem-water dynamics
21. Socio-economic drivers	Socio-economic factors influencing water management decisions and behaviours
22. Governance mechanisms	Institutional frameworks and policy instruments governing water resources and ecosystem services
23. Technological innovations	Innovative solutions and technologies for enhancing ecosystem-water resilience
24. Capacity building	Capacity-building initiatives to empower stakeholders and promote ecosystem-based water management practices
25. Knowledge sharing	Platforms and mechanisms for sharing knowledge and best practices in ecosystem-based water management
26. Ecological processes	Understanding of ecological processes influencing water quantity, quality, and resilience
27. Hydrological interactions	Dynamics of water resources, hydrological processes, and their interactions with ecosystems
28. Social equity	Considerations of equity and justice in water resource allocation and access
29. Financing mechanisms	Innovative financing mechanisms for supporting ecosystem conservation and restoration efforts
30. Scaling and replication	Scalability and replicability of successful ecosystem-based solutions across different contexts and regions
31. Cultural considerations	Cultural, social, and institutional factors influencing ecosystem-water interactions and management practices

Source Created by authors

vital services to humans. Recognising and incorporating ecosystem-based methods into water resource management strategies ensures hydrological system durability and viability despite environmental and human impacts. Natural systems, such as wetlands, rivers, forests, and oceans, play a crucial role in purifying water through biological, physical, and chemical mechanisms. Wetlands filter pollutants, retain sediment, and detoxify harmful substances, providing clean, safe water for humans and the environment. They capture and keep contaminants and nutrients in their soils by settling suspended particles and sediments from water bodies (Bogachev et al. 2023; Hong et al. 2024; Ranjan 2024; Rezvani et al. 2024). Wetland plants and microbes absorb and assimilate nutrients from the water column, reducing eutrophication risks in downstream ecological systems. River and stream ecosystems also play a vital role in water purification. They help dilution, dispersion, sedimentation, and filtration of pollutants, aiding nutrient cycling and biological uptake. Riverine habitat preservation and rehabilitation are essential for water quality, biodiversity, and aquatic system ecological integrity. Forests naturally filter sediments, nutrients, and pollutants from surface runoff and groundwater, facilitating water infiltration through physical and chemical processes. Forests also regulate streamflow and maintain baseflow, protecting watersheds, replenishing groundwater, and improving water quality. Ocean and coastal ecosystems use biological, physical, and chemical mechanisms to break down pollutants and contaminants. Ocean circulation patterns help disperse and dilute pollutants, while phytoplankton blooms remove carbon and nutrients from the water column, improving water clarity and quality. Marine bacteria, algae, and invertebrates biodegrade and detoxify pollutants and contaminants in marine and estuarine environments. Coastal ecosystems filter sediments, nutrients, and pollutants from coastal runoff, decompose organic matter and hazardous substances, protect cycle nutrients, support fisheries, and meet cultural and recreational needs. Integrating ecological methods into water resource management strategies is crucial for sustainability and aquatic environment health. We must address water quality issues and improve marine ecosystems' resilience to environmental and human impacts (Fig. 7.2).

Fig. 7.2 Ecosystem services for water resources. *Source* Created by authors

Role of Biodiversity in Maintaining Ecosystem Resilience and Water Quality

Nature uses biological, physical, and chemical mechanisms to purify water, providing clean, safe water for humans and the environment. Wetlands, rivers, forests, and oceans filter pollutants, retain sediment, and detoxify harmful substances, resulting in clean, safe water for humans and the environment. Wetlands capture and keep contaminants and nutrients by settling suspended particles and sediments from water bodies, while rivers and streams help purify marine water through physical, chemical, and biological mechanisms. Riverine habitat preservation and rehabilitation are essential for water quality, biodiversity, and aquatic system ecological integrity. Forests naturally filter sediments, nutrients, and pollutants from surface runoff and groundwater, helping soil and vegetation absorb and retain water. Forests regulate streamflow and maintain baseflow, protecting watersheds, replenishing groundwater, and improving water quality. Oceans and coastal ecosystems, mainly marine and estuarine, break down pollutants and contaminants using biological, physical, and chemical mechanisms (Qin et al. 2021; Shi et al. 2023; Zhu et al. 2023; Miao et al. 2024; Zhang et al. 2024). Oceanic processes transport water masses and suspended particles over long distances, mixing and diffusing contaminants and pollutants. Phytoplankton blooms remove carbon and nutrients from the water column, improving water clarity and quality. Bacteria, algae, and invertebrates help biodegrade and detoxify pollutants and contaminants in marine and estuarine environments. Coastal ecosystems filter sediments, nutrients, and pollutants from coastal runoff, decompose organic matter and hazardous substances, and support fisheries. Marine ecosystems use biological, physical, and chemical processes to remove and break down pollutants and contaminants, making them vital to water purification. Preservation and rehabilitation of these ecological systems are crucial for water resource sustainability and aquatic environment health. We must address water quality issues and improve marine ecosystems' resilience to environmental and human impacts (Fig. 7.3).

Climate Change Impacts on Water Resources

Climate change transforms rainfall and extreme hydrological events, impacting water accessibility, quality, and management worldwide. It has increased rainfall, making many areas more susceptible to flash floods and urban flooding. Climate change also alters seasonal rainfall patterns, causing more prolonged droughts, heavier rainfall in some regions, and increased rainfall in others. Regional differences in water accessibility and hydrological patterns result from atmospheric circulation patterns like jet stream and monsoon system relocation. Climate change intensifies hydrological extremes like droughts, floods, and storms, affecting water resources, organisms, and human communities. These extremes cause water scarcity, agricultural losses, infrastructure damage, and public health risks. Climate change worsens

Fig. 7.3 Role of biodiversity in maintaining ecosystem resilience and water quality. *Source* Created by authors

droughts, causing water scarcity, crop failures, and ecological damage. It can increase precipitation and runoff, increasing flood frequency and scale in river and coastline regions. Sea level rise, storm surges, and intense precipitation increase flood risks, submerging low-lying areas, damaging infrastructure, and displacing communities. Landslides, mudslides, and debris flows from floods threaten human and ecological safety. Climate change also alters the frequency and intensity of tropical cyclones, hurricanes, and typhoons, affecting water resources and coastal environments. Storms, high winds, and heavy rainfall increase flood risks and infrastructure damage, making disaster preparedness and response difficult. Water managers must predict and address rainfall patterns, extreme hydrological events, and water needs, including water conservation, drought readiness, flood management, and ecological restoration. Sustainable water management requires collaborative governance, stakeholder engagement, and science-based decision-making (Fig. 7.4). Climate change poses significant challenges to global ecosystems, economies, and societies, including water scarcity, water quality degradation, and competition for scarce water resources. Comprehensive water management strategies must be implemented to mitigate these effects, including water conservation, pollution reduction, ecological restoration, and climate resilience (Belete 2023; Perera and Hong 2023; Abdi et al. 2024; Chung et al. 2024; Huang et al. 2024).

Fig. 7.4 Climate change impacts on water resources. *Source* Created by authors

Ecosystem Resilience and Water Management

Ecosystem-based water resource management is a sustainable approach that integrates ecological principles, processes, and services into decision-making. This approach enhances water system and ecosystem resilience, sustainability, and adaptability. It recognises human-natural interdependence and promotes environmental preservation, rehabilitation, and sustainable use to address water issues and advance socioeconomic development. Ecosystem-based watershed management prioritises watershed protection, rehabilitation, and sustainability. Riparian zones along rivers and streams provide ecological services like sediment retention, nutrient cycling, and habitats. Restoration projects increase riparian vegetation, restore hydrological mechanisms, and improve aquatic habitats. Forest-covered watersheds regulate hydrological cycles, stabilise soil, and mitigate erosion, preserving water quality and quantity. Wetland conservation strategies aim to protect biodiversity, improve hydrological system connectivity, and reduce land use and climate change's adverse effects on wetland ecosystems. Ecosystemic green infrastructure manages water resources, reduces urban runoff, and improves urban water quality using natural methods. Green infrastructure strategies include urban parks, green roofs, and vegetated buffers, which help manage water and reduce pollution. Ecological services in water infrastructure planning involve understanding ecosystems' effects on water resources, identifying human-natural system compromises and mutual benefits, and incorporating ecological factors into decision-making (Dodd et al. 2023; Karim et al. 2023; Yang et al. 2023; James et al. 2024). Evaluating ecosystem biodiversity services helps prioritise conservation and restoration efforts, identify vulnerable habitats and species that need conservation, and allow ecological factors to be considered in water infrastructure planning. Stakeholder engagement promotes agreement, cooperation, and inclusive decision-making that considers all stakeholders. Adaptive management strategies account for socio-ecological systems' uncertainty, complexity, and dynamic nature. Ecological services help monitor ecological changes, assess management goals, and identify opportunities for adaptive management and ongoing enhancement. Collaboration, stakeholder engagement, and adaptive management are essential for balancing human needs and environmental conservation goals, promoting environmental well-being and resilience, and ensuring future water resource sustainability (Fig. 7.5).

Fig. 7.5 Ecosystem resilience and water management. *Source* Created by authors

Ecosystem-Based Water Resource Management Initiatives

Ecosystem-based water resource management initiatives focus on conserving, restoring, and promoting sustainable use of natural ecosystems to address water issues, promote socio-economic development, and improve ecosystem health and resilience. These strategies integrate ecological principles into water management and infrastructure planning decisions, supporting biodiversity conservation, water resource sustainability, and climate change adaptation. They include river and stream riparian zones, forested watersheds, and wetlands, which support biodiversity, water quality, and habitat provision while preventing deforestation and promoting sustainable land management. To improve wetland resilience and biodiversity, we remove invasive species, reestablish native vegetation, and recreate natural hydrological regimes (Angeler et al. 2023; Wang et al. 2024). Natural infrastructure like wetlands manages water and mitigates floods cheaply. Wetland ecosystem conservation and restoration can protect against floods, purify water, sequester carbon, and provide recreational opportunities while strengthening ecosystems and supporting climate adaptation (Gogoi et al. 2024). Green infrastructure, such as urban parks, green roofs, and vegetated buffers, manages water resources, reduces urban runoff, and improves water quality using natural methods. Engineered wetlands mimic natural wetlands' water filtration, nutrient cycling, and habitat services, supporting aquatic habitats and recreation. Coastal ecosystems like mangroves, salt marshes, and coral reefs require management to address coastal hazards, habitat degradation, and pollution threats. Integrated water resources management (IWRM) integrates water planning, allocation, and management to achieve sustainable development. Ecosystem-based water resource management requires diverse stakeholders to participate in decision-making, fostering consensus and collaboration. Adaptive management uses ecosystem monitoring, data collection, and knowledge sharing to promote ecosystem health, human well-being, and socio-economic development. Science-based decision-making ensures water management decisions are based on sound ecological principles, best science, and stakeholder input. These initiatives create innovative, cost-effective solutions to water issues, resilience, and sustainable development, protecting water resources, conserving biodiversity, and strengthening ecosystems and human communities. Ecosystem-based approaches and water resource sustainability for future generations require collaborative governance, stakeholder engagement, and adaptive management (Fig. 7.6).

Policy and Governance

Ecological resilience in water management strategies is crucial to sustainable water use. It involves legal, institutional, and regulatory structures to promote ecosystem conservation, restoration, and sustainable use. These strategies include laws, regulations, guidelines, incentives, and decision-support tools. Legal frameworks define

Fig. 7.6 Ecosystem-based water resource management initiatives. *Source* Created by authors

stakeholders' rights, regulate water use, and enforce environmental laws. Institutional mechanisms provide the organisational structure and governance needed to implement these strategies. Government agencies, intergovernmental organisations, civil society groups, and private sector actors collaborate to promote sustainable water management and ecological conservation. Pollution control regulations, water pricing mechanisms, environmental restoration incentives, and market-based instruments help integrate ecological principles into water management strategies. Stakeholder involvement is essential for addressing water issues and promoting sustainable development for future generations. Indigenous communities contribute significantly to environmental water resilience solutions using traditional knowledge and practices to manage water, land, and biodiversity. Community-based initiatives empower residents and promote sustainable water management. Academic and research institutions are vital in developing ecologically-based water resilience solutions (Retallack 2021; Liu et al. 2022; Liang and Li 2023).

Innovations and Future Directions

Emerging technologies and methods enhance ecosystem and water system resilience by enabling innovative monitoring, modelling, and management. These technologies integrate ecological principles, processes, and services into decision-making, supporting environmental conservation and restoration. Satellite imagery, aerial photography, and uncrewed aerial vehicles (UAVs) can monitor ecological well-being, water resource quality, and land cover changes over large areas and time-frames. Remote sensing data can map wetlands, forests, and riparian areas, monitor vegetation cover, identify land use changes, and assess water quality parameters like turbidity, chlorophyll concentration, and temperature. Geographic Information Systems (GIS) are powerful tools for spatial analysis, mapping, and visualising landscape-water interactions. These systems integrate remote sensing, field surveys, and socioeconomic indicator data to create spatially explicit models and decision-support systems. GIS tools help identify critical ecological system preservation and rehabilitation areas, evaluate ecological services, and inform land use planning and water management strategies to improve the ecological system and water resource resilience. Machine learning, neural networks, and deep learning algorithms can analyse complex datasets, forecast ecological reactions, and optimise management strategies to strengthen water systems. AI monitors water quality, models hydrology, and forecasts ecology more accurately and effectively using data science, computational modelling, and predictive analytics (Bao et al. 2022; Dias et al. 2023; Macedo et al. 2023; Marcos et al. 2023; Delgado et al. 2024). Neural network models, such as ANNs, CNNs, and RNNs, can predict hydrological variables, simulate ecological system dynamics, and assess climate change's effects on water resources. In-situ, remote, and IoT sensors can monitor and collect real-time data in natural and built environments and evaluate water quality, quantity, and the health of natural and built environments. Connecting intelligent sensors, data loggers, and environmental monitoring stations to networks and platforms allows for scalable and cost-effective water quality, quantity, and ecological well-being monitoring. Ecohydrological modelling simulates environmental and water dynamics, predicting future scenarios and evaluating management options to improve environmental resilience and sustainable water management. These models help decision-makers assess ecological services, plan land use, and manage water. Process-based models replicate the physical, chemical, and biological processes that regulate ecosystem and water interactions (Wollstein et al. 2022; Anamaghi et al. 2023; Dowlati et al. 2023; Behboudian et al. 2024). Water management uses process-based models to represent ecosystem dynamics and predict ecological responses to environmental change. These models help stakeholders evaluate trade-offs, prioritise interventions, and optimise management strategies to improve ecosystem resilience and water security. Scenario analysis uses ecohydrological models to evaluate alternative management options, investigate future situations, and identify intervention opportunities. However, several research areas remain to improve the understanding and application of ecosystem-based water management. These areas span various disciplines, including hydrological interactions, socio-economic drivers, and governance mechanisms. Researchers, practitioners, policymakers, and stakeholders must collaborate to solve complex water issues and achieve sustainable development goals. Research on ecological processes, nutrient cycling, species interactions, and ecosystem services is crucial for understanding water quantity, quality, and resilience. Understanding the institutional, legal, and policy structures that regulate water resources, ecosystem services, and water management practices is also essential. Integrating ecosystem-based approaches into water governance is crucial for promoting collaborative decision-making, adaptive management, and sustainable development. Ecosystem-based water management research gaps and advances require collaboration across academic domains, stakeholder engagement, and adaptive management strategies that incorporate ecological, hydrological, socio-economic, and governance perspectives into decision-making (Xue et al. 2020; Gillson et al. 2021; Zhao et al. 2021; Huang et al. 2023; Lei et al. 2023; Suzette Lorilla et al. 2023; Wang et al. 2023; Bekri et al. 2024; Han 2024).

Conclusion

In summary, the complex relationship between the resilience of ecosystems and the availability of water resources highlights the significance of conserving and rehabilitating natural ecological systems to ensure sustainable water management in light of increasing environmental difficulties. We have explained various complex aspects of this mutually beneficial association, encompassing the regulatory functions of forests, wetlands, and rivers in hydrological cycles and the crucial role of biodiversity in preserving water quality. The chapter explanation indicate that ecosystems play an active role in mitigating the impacts of climate change rather than being passive recipients. They serve as buffers against adverse effects and provide essential ecological services that protect water availability, regulate water quality, and alleviate hydrological extremes. Nevertheless, as our comprehension of these interrelationships becomes more profound, the imperative to preserve and restore these ecological systems also intensifies. Acknowledging the crucial significance of stakeholders, encompassing governmental bodies, non-governmental organisations, local communities, and indigenous groups, is of utmost importance in successfully implementing ecologically-based approaches to enhance water resilience. To facilitate collective action and advance the incorporation of ecological considerations into water management strategies, it is imperative to establish collaborative governance mechanisms, inclusive decision-making processes, and capacity-building initiatives. Furthermore, emerging technologies and innovative methodologies show the potential to augment our capacity to observe, simulate, and regulate the interactions between ecosystems and water. This offers valuable resources for making well-informed decisions and implementing adaptive management strategies in unpredictable circumstances. Through the utilisation of remote sensing, artificial intelligence, sensor technologies, and eco-hydrological modelling, the potential exists to unlock novel prospects for the progression of ecosystem-based water management and the resolution of intricate water-related issues. Nevertheless, it is crucial to recognise and tackle the existing research deficiencies and unresolved issues in pursuing a more sustainable way for future. There is an urgent requirement for interdisciplinary collaboration, knowledge exchange, and policy innovation to fully harness ecological resilience's potential in safeguarding water resources for present and future generations. This includes comprehending ecological processes and hydrological interactions and promoting stakeholder engagement and capacity-building initiatives. Fundamentally, the pursuit of water resilience is inherently interconnected with the conservation and rehabilitation of robust ecological systems. By adopting ecological-based methodologies, utilising technological advancements, and promoting inclusive governance, it is possible to effectively address the intricate challenges posed by a shifting climate and environment while ensuring the preservation of the essential water resources that support life on our planet. The resilience and vitality of ecosystems and the water resources they provide for the well-being of humanity and the earth can only be ensured through concerted action and collective stewardship.

References

Abdi S, Yazdani M, Najafi E (2024) Evaluating innovation ecosystem resiliency using agent-based modeling and systems dynamics. J Saf Sci Resil. https://doi.org/10.1016/j.jnlssr.2024.03.004

Anamaghi S, Behboudian M, Mahjouri N, Kerachian R (2023) A resilience-based framework for evaluating the carrying capacity of water and environmental resources under the climate change. Sci Total Environ 902:165986. https://doi.org/10.1016/j.scitotenv.2023.165986

Angeler DG, Heino J, Rubio-Ríos J, Casas JJ (2023) Connecting distinct realms along multiple dimensions: a meta-ecosystem resilience perspective. Sci Total Environ 889:164169. https://doi.org/10.1016/j.scitotenv.2023.164169

Bao Z, Shifaw E, Deng C et al (2022) Remote sensing-based assessment of ecosystem health by optimizing vigor-organization-resilience model: a case study in Fuzhou City, China. Ecol Inform 72:101889. https://doi.org/10.1016/j.ecoinf.2022.101889

Behboudian M, Anamaghi S, Kerachian R, Kalantari Z (2024) Comparison of three group decision-making frameworks for evaluating resilience time series of water resources systems under uncertainty. Ecol Indic 158:111269. https://doi.org/10.1016/j.ecolind.2023.111269

Bekri ES, Kokkoris IP, Skuras D, et al (2024) Ecosystem accounting for water resources at the catchment scale, a case study for the Peloponnisos, Greece. Ecosyst Serv 65:101586. https://doi.org/10.1016/j.ecoser.2023.101586

Belete MD (2023) Ecohydrological nature-based solution for climate resilience and biodiversity enhancement in water-limited ecosystem: perspectives and proof of concepts. Ecohydrol Hydrobiol 23:507–517. https://doi.org/10.1016/j.ecohyd.2023.08.011

Bogachev MI, Gafurov AM, Iskandirov PY et al (2023) Reversal in the drought stress response of the Scots pine forest ecosystem: local soil water regime as a key to improving climate change resilience. Heliyon 9:e21574. https://doi.org/10.1016/j.heliyon.2023.e21574

Choudhury M, Sharma A, Singh P, Kumar D (2021) Impact of climate change on wetlands, concerning Son Beel, the largest wetland of North East, India. In: Global climate change. Elsevier, pp 393–414. https://doi.org/10.1016/B978-0-12-822928-6.00006-X

Choudhury M, Pervez A, Sharma A, Mehta J (2022) Human-induced stresses on the rivers beyond their assimilation and regeneration capacity. In: Ecological significance of river ecosystems. Elsevier, pp 281–298. https://doi.org/10.1016/B978-0-323-85045-2.00019-4

Chung HI, Kim Y, Ryu J et al (2024) Exploring a novel methodology for post impact assessments of chemical accidents on forest ecosystems: with damage, recovery, and resilience aspects of vegetation. Int J Disaster Risk Reduct 100:104059. https://doi.org/10.1016/j.ijdrr.2023.104059

Delgado A, Rodriguez DJ, Amadei CA, Makino M (2024) Water in circular economy and resilience (WICER) framework. Util Policy 87:101604. https://doi.org/10.1016/j.jup.2023.101604

Dias HQ, Sukumaran S, Neetu S et al (2023) Deciphering benthic ecosystem functioning and resilience in a major port and marine protected area via the multi-trait approach. Environ Res 237:116988. https://doi.org/10.1016/j.envres.2023.116988

Dodd RJ, Chadwick DR, Hill PW et al (2023) Resilience of ecosystem service delivery in grasslands in response to single and compound extreme weather events. Sci Total Environ 861:160660. https://doi.org/10.1016/j.scitotenv.2022.160660

Dowlati M, Seyedin H, Behnami A et al (2023) Water resources resilience model in climate changes with community health approach: qualitative study. Case Stud Chem Environ Eng 8:100521. https://doi.org/10.1016/j.cscee.2023.100521

Gillson L, Dirk C, Gell P (2021) Using long-term data to inform a decision pathway for restoration of ecosystem resilience. Anthropocene 36:100315. https://doi.org/10.1016/j.ancene.2021.100315

Gogoi N, Choudhury M, Ul Hasan MS, Changmai B, Baruah D, Samanta P (2024) Soil organic carbon pool in diverse land utilization patterns in North-East India: an implication for carbon sequestration. Environ Dev Sustain, 1–27. https://doi.org/10.1007/s10668-024-05037-y

Han S (2024) Integrating mineral and natural resources for enhanced environmental resilience and sustainability. Resour Policy 91:104869. https://doi.org/10.1016/j.resourpol.2024.104869

Hong G, Liu S, Liu W, Wu X (2024) Nonlinear trade-off relationship and critical threshold between ecosystem services and climate resilience for sustainable urban development. Sustain Cities Soc 103:105253. https://doi.org/10.1016/j.scs.2024.105253

Huang H, Zhuo L, Wang W, Wu P (2023) Resilience assessment of blue and green water resources for staple crop production in China. Agric Water Manag 288:108485. https://doi.org/10.1016/j.agwat.2023.108485

Huang S-W, Hussain B, Chen J-S et al (2024) Evaluating groundwater ecosystem dynamics in response to post in-situ remediation of mixed chlorinated volatile organic compounds (CVOCs): an insight into microbial community resilience, adaptability, and metabolic functionality for sustainable remediatio. Sci Total Environ 920:170874. https://doi.org/10.1016/j.scitotenv.2024.170874

James RK, Daniels F, Chauhan A et al (2024) Monitoring vegetation patterns and their drivers to infer resilience: Automated detection of vegetation and megaherbivores from drone imagery using deep learning. Ecol Inform 81:102580. https://doi.org/10.1016/j.ecoinf.2024.102580

Karim MR, Mukul SA, Zahir RB et al (2023) The role of protected areas co-management in enhancing resistance and resilience of deciduous forest ecosystem to extreme climatic events in Bangladesh. J Environ Manage 326:116800. https://doi.org/10.1016/j.jenvman.2022.116800

Kumar D, Choudhury M, Rathore A (2020) Valuation of ecosystem services & benefits of Son Beel wetland in Assam, India: a case study of natural solutions to climate change & water: the worldwide online climate change conference. Hamburg University, Germany

Kumar D, Choudhury M (2021) Recognizing economic values of wetland ecosystem services: a study of emerging role of monetary evaluation of Chandubi ecosystem and biodiversity. In: Wetlands conservation: current challenges and future strategies, pp 87–110. https://doi.org/10.1002/9781119692621.ch5

Lei J, Li C, Yang W (2023) Ecosystem health assessment and approaches to improve Sichuan Province based on an improved vigor organization resilience model. Ecol Indic 155:110925. https://doi.org/10.1016/j.ecolind.2023.110925

Liang L, Li Y (2023) How does government support promote digital economy development in China? The mediating role of regional innovation ecosystem resilience. Technol Forecast Soc Change 188:122328. https://doi.org/10.1016/j.techfore.2023.122328

Liu B, Han R, Tang L, Cheng X (2022) An object–pressure–endurance–resilience system for assessing the vulnerability of coastal zone ecosystems. Ecol Indic 141:109105. https://doi.org/10.1016/j.ecolind.2022.109105

Macedo M, Fabré NN, da Silva VEL et al (2023) Influence of the river flow pulse on the maturity, resilience, and sustainability of tropical coastal ecosystems. Mar Environ Res 183:105806. https://doi.org/10.1016/j.marenvres.2022.105806

Marcos B, Gonçalves J, Alcaraz-Segura D et al (2023) Assessing the resilience of ecosystem functioning to wildfires using satellite-derived metrics of post-fire trajectories. Remote Sens Environ 286:113441. https://doi.org/10.1016/j.rse.2022.113441

Miao X, Li Z, Wang M et al (2024) Measurement of cultivated land ecosystem resilience in black soil region of Northeast China under the background of cultivated land protection policy in China: case study of Qiqihar City. J Clean Prod 434:140141. https://doi.org/10.1016/j.jclepro.2023.140141

Perera ATD, Hong T (2023) Vulnerability and resilience of urban energy ecosystems to extreme climate events: a systematic review and perspectives. Renew Sustain Energy Rev 173:113038. https://doi.org/10.1016/j.rser.2022.113038

Qin J, Hao X, Hua D, Hao H (2021) Assessment of ecosystem resilience in Central Asia. J Arid Environ 195:104625. https://doi.org/10.1016/j.jaridenv.2021.104625

Ranjan R (2024) Creating synergies between payments for ecosystem services, green bonds, and catastrophe insurance markets for enhanced environmental resilience. Land use policy 136:106970. https://doi.org/10.1016/j.landusepol.2023.106970

Retallack M (2021) The intersection of economic demand for ecosystem services and public policy: a watershed case study exploring implications for social-ecological resilience. Ecosyst Serv 50:101322. https://doi.org/10.1016/j.ecoser.2021.101322

Rezvani SMHS, Silva MJF, de Almeida NM (2024) NETOBRA: boosting urban resilience through a digital platform for the construction ecosystem. Procedia Struct Integr 55:72–79. https://doi.org/10.1016/j.prostr.2024.02.010

Sharma A, Choudhury M (2021) Changing climate and depleting water resources in the mountains with a case study from the Himalayas. In: Water conservation in the era of global climate change. Elsevier, pp 411–436. https://doi.org/10.1016/B978-0-12-820200-5.00013-0

Shi X, Zhang Y, Ding H et al (2023) Drought risk assessment considering ecosystem resilience: a case study in the Huang-Huai-Hai Plain, China. Ecol Indic 156:111102. https://doi.org/10.1016/j.ecolind.2023.111102

Suzette Lorilla R, Kefalas G, Christou AK et al (2023) Enhancing the conservation status and resilience of a narrowly distributed forest: a challenge to effectively support ecosystem services in practice. J Nat Conserv 73:126414. https://doi.org/10.1016/j.jnc.2023.126414

Wang F, Wong W-K, Wang Z et al (2023) Emerging pathways to sustainable economic development: an interdisciplinary exploration of resource efficiency, technological innovation, and ecosystem resilience in resource-rich regions. Resour Policy 85:103747. https://doi.org/10.1016/j.resourpol.2023.103747

Wang Y, Wan F, Peng S et al (2024) Resilience level assessment of regional water resources system based on multi-link. Ecol Indic 160:111943. https://doi.org/10.1016/j.ecolind.2024.111943

Wollstein K, Creutzburg MK, Dunn C et al (2022) Toward integrated fire management to promote ecosystem resilience. Rangelands 44:227–234. https://doi.org/10.1016/j.rala.2022.01.001

Xue B, Wang G, Xiao J et al (2020) Global convergence but regional disparity in the hydrological resilience of ecosystems and watersheds to drought. J Hydrol 591:125589. https://doi.org/10.1016/j.jhydrol.2020.125589

Yang S, Song S, Li F et al (2023) Evaluating farmland ecosystem resilience and its obstacle factors in Ethiopia. Ecol Indic 146:109900. https://doi.org/10.1016/j.ecolind.2023.109900

Zhang R-Q, Xiong Q, Wu L et al (2024) Water supply following drought: effects on drougt legacy and resilience in a tropical forest—a case study in Xishuangbanna, China. Ecol Inform 79:102422. https://doi.org/10.1016/j.ecoinf.2023.102422

Zhao R, Fang C, Liu H, Liu X (2021) Evaluating urban ecosystem resilience using the DPSIR framework and the ENA model: a case study of 35 cities in China. Sustain Cities Soc 72:102997. https://doi.org/10.1016/j.scs.2021.102997

Zhu S, Li D, Feng H, Zhang N (2023) The influencing factors and mechanisms for urban flood resilience in China: From the perspective of social-economic-natural complex ecosystem. Ecol Indic 147:109959. https://doi.org/10.1016/j.ecolind.2023.109959

Chapter 8
Safeguarding Aquatic Ecosystems: Innovations in Water Resource Management

Sukhendu Dey, Kamalesh Sen, Shubhajit Saha, Nimai Chandra Saha, Daniel D. Snow, and Md. Yeamin Hossain

Abstract Sustainable development necessitates the comprehensive management of water resources in light of diminishing supply and increasing demand. This review examines new developments in methods, instruments, and policies to deal with issues brought on by population expansion, urbanization, and climate change. It starts by looking at water conservation techniques, such as leak detection devices, intelligent irrigation systems, and rainwater harvesting, which maximize home, commercial, and agricultural usage and support water sustainability. It discusses advanced technologies in wastewater treatment, like membrane filtration and decentralized systems, improving water quality and enabling the reuse of treated wastewater. This review highlights innovative approaches to governance and policy, such as integrated frameworks and stakeholder engagement strategies, ensuring equitable resource allocation. Additionally, it investigates how artificial intelligence, remote sensing, and data analytics may improve water management techniques by enabling real-time

---

S. Dey
The University of Burdwan, Burdwan, West Bengal 713104, India

K. Sen (✉)
Department of Environmental Science, Syamsundar College, Shyamsundar, Burdwan, West Bengal 713424, India
e-mail: ksen.envs@gmail.com

S. Saha
Fishery and Ecotoxicology Research Laboratory, Department of Zoology, The University of Burdwan, Burdwan, India

N. C. Saha
Post Graduate Department of Zoology, Bidhannagar College, Sector 1, Bidhannagar, Kolkata, West Bengal 700064, India

D. D. Snow
School of Natural Resources and Nebraska Water Center, part of the Daugherty Water for Food Global Institute, University of Nebraska, Lincoln, NE 68583-0844, USA

Md. Y. Hossain
Department of Fisheries, Faculty of Fisheries, University of Rajshahi, Rajshahi 6205, Bangladesh

M. Choudhury et al. (eds.), *Sustainable Synergy: Harnessing Ecosystems for Climate Resilience*, Climate Change Management,
https://doi.org/10.1007/978-3-031-77957-2_8

monitoring and predictive modelling for well-informed decision-making. Integration and innovation are crucial for addressing water scarcity, pollution, and access disparities and building resilient water systems for present and future generations.

Keywords Water resource management · Sustainability · Conservation · Treatment · Stakeholder engagement

Introduction

Given the growing demand and decreasing water supply, sustainable development necessitates a comprehensive strategy for water resource management (Inyinbor Adejumoke et al. 2018; Bhatt et al. 2022; Al-Hazmi et al. 2023). This chapter explores recent advances in water resource management strategies, tools, and regulations that aim to address challenges arising from urbanization, population growth, and climate change (Inyinbor Adejumoke et al. 2018; Al-Hazmi et al. 2023). As urban areas continue to expand, the pressure on water resources intensifies (Liquete et al. 2016). Urbanization leads to increased water demand for domestic, industrial, and commercial purposes (Wang et al. 2021). Moreover, urban development often results in water quality degradation through pollution from runoff, industrial discharges, and sewage (Müller et al. 2020). To mitigate these challenges, innovative water management strategies are essential. This includes implementing green infrastructure such as permeable pavements, rain gardens, and green roofs to manage stormwater runoff effectively (Zavala-Yoe et al. 2020). Finally, municipalities may identify and address water losses in urban infrastructure using smart water metering and leak detection systems, thereby conserving water resources. Finally, municipalities may identify and address water losses in urban infrastructure using smart water metering and leak detection systems, thereby conserving water resources (Cosgrove and Loucks 2015; Majumder and Kaviraj 2017).

The pressure on water resources is increased by population growth, especially in areas where water shortage already exists (McGrane 2016). As the global population continues to raise, the demand for water for drinking, sanitation, agriculture, and energy production increases correspondingly (Nestler et al. 2011). Effective water management strategies and creative water conservation approaches are required to fulfil the expanding demand sustainably (Rauh and Hughes 2024; Abdelfattah and El-Shamy 2024). This includes the promotion of water-saving technologies in households and industries, as well as the adoption of precision agriculture methods to optimize water use in farming (Liquete et al. 2016). Additionally, neighborhood-based water management programs that include local stakeholders in decision-making procedures can improve water governance and encourage fair distribution of water resources (Martinho 2019).

Climate change poses significant challenges to water resource management, leading to altered precipitation patterns, increased frequency of extreme weather events (Saleem et al. 2024), and rising sea levels (Inyinbor Adejumoke et al. 2018;

Borzì 2023). These changes impact water availability, quality, and distribution, exacerbating water stress in many regions (Zavala-Yoe et al. 2020). Building resilience in the face of water issues brought on by climate change requires implementing adaptation techniques (Saleem et al. 2024). Adaption can include the development of climate-resilient water infrastructure, such as resilient water supply systems and flood management infrastructure (Duchenne-Moutien and Neetoo 2021). Furthermore, by increasing water storage capacity and promoting water quality, nature-based solutions like wetland restoration and watershed management can help mitigate the effects of climate change on water supplies (Liquete et al. 2016; Kumar and Choudhury 2021). Surface water storage has several advantages, such as providing a visible pool of water in surface water reservoirs, making it easier to manage than groundwater storage. However, there are many drawbacks to surface water storage as well, including the requirement for substantial surface areas, the possibility of evaporation, and the influence of human activity on water quality (Fig. 8.1).

In response to these challenges, there have been significant advancements in water resource management strategies, tools, and regulations (Zavala-Yoe et al. 2020). Innovations such as smart irrigation systems, leak detection technologies, and rainwater harvesting methods help optimize water usage in agricultural, industrial, and domestic settings, contributing to water sustainability (Liquete et al. 2016). Emerging technologies in wastewater treatment, such as membrane filtration and advanced oxidation processes, improve water quality and facilitate the reuse of treated wastewater for non-potable purposes, reducing the strain on freshwater resources (Armah et al. 2020; Rajpal et al. 2022). Innovative approaches to water governance, such as Integrated Water Resources Management (IWRM) frameworks and stakeholder

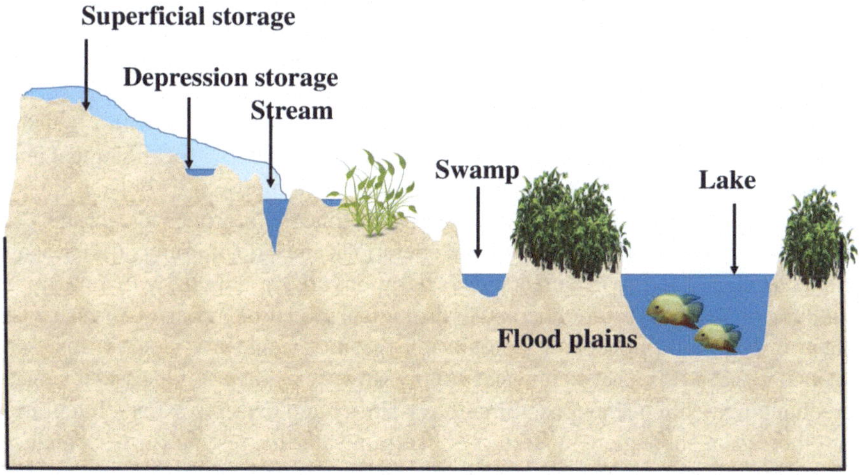

Fig. 8.1 Surface water storage includes water on the soil surface, water retained as crops, and water held in depressions such as floodplains, lakes, wetlands, and pools

engagement strategies, promote inclusive procedures for making decisions and guarantee a fair allocation of water resources (Zavala-Yoe et al. 2020). The integration of data analytics, remote sensing, and artificial intelligence enables real-time monitoring of water quality and quantity, predictive modelling of water availability, and informed decision-making, empowering stakeholders to respond effectively to dynamic water resource issues (Hauser-Davis and Wosnick 2022; Borzì 2023).

This review is a comprehensive resource that provides a deep understanding of how recent advancements in water resource management strategies, tools, and regulations can address the challenges of urbanization, population growth, and climate change. By leveraging these innovations, societies can significantly enhance water sustainability, improve water quality, and ensure equitable access to water supplies for the current and upcoming generations.

Aquifer Storage and Recharge (ASR)

ASR marks a pioneering stride in water resources management (Maliva 2014), providing a sustainable avenue for storing surplus water during abundance and replenishing depleted aquifers during scarcity (Bandyopadhyay et al. 2021). By purposefully injecting water into underground aquifers for later use, ASR becomes instrumental in mitigating drought impacts, lessening reliance on surface water reservoirs, and preventing saltwater intrusion in coastal regions (Prusty and Farooq 2020). Employing diverse injection techniques like wells, infiltration basins, or recharge ponds, ASR systems introduce surplus surface water or treated wastewater into underground aquifers (Samal et al. 2020), with advanced methods like managed aquifer recharge meticulously optimizing storage capacity and water quality (Table 8.1).

Recovery methods range from pumping from dedicated wells to utilizing groundwater flow towards extraction points. Advanced water treatment processes such as filtration, disinfection, and chemical treatment ensure compliance with regulatory standards and maintain aquifer water quality (Samal et al. 2020). Technological innovations like membrane filtration and advanced oxidation processes bolster the efficiency and reliability of ASR operations, with real-time monitoring and data analytics playing a crucial role in optimizing ASR operations and safeguarding aquifer health (Othman et al. 2022). Environmental considerations are paramount, with innovations in monitoring, habitat restoration, and hydrological modeling aiding in assessing and mitigating potential impacts on groundwater resources and ecosystems (Prusty and Farooq 2020). ASR is a resilient water supply strategy by diversifying water sources and enhancing resilience against climate variability and extreme events. Integration with other water management strategies like rainwater harvesting further enhances resilience to climate change impacts (Bandyopadhyay et al. 2021). Successful ASR implementation necessitates supportive regulatory frameworks, stakeholder collaboration, and public engagement facilitated by innovations in policy development and transparent communication. ASR represents groundbreaking approaches to water resources management, offering sustainable solutions for water storage, supply

Table 8.1 Overview of various ASR methods

Type of ASR	Hydrogeological management	References
Well Injection	Involves injecting water directly into underground aquifers through wells	Alley (2009)
Infiltration Basins	Utilizes constructed basins or depressions to facilitate the infiltration of water into aquifers	Sasidharan et al. (2021)
Recharge Ponds	Artificial ponds or reservoirs designed to hold water for infiltration into aquifers	Deng and Bailey (2022)
Managed Recharge	Controlled and optimized injection of water into aquifers to maximize storage capacity and maintain water quality	Casanova et al. (2016)
Indirect Recharge	Channels excess surface water or treated wastewater to areas where natural recharge can occur	Shawaqfah et al. (2021)
Enhanced Recharge	Augments natural recharge processes through engineering interventions such as recharge wells or spreading basins	Shawaqfah et al. (2021)
Controlled Recharge	Regulates the rate and timing of water injection to balance aquifer storage and withdrawal	Amini and Homayounfar (2017)
Natural Infiltration	Relies on natural processes such as rainfall or runoff to recharge aquifers without human intervention	Amini and Homayounfar (2017)

augmentation, and groundwater replenishment (Bergström et al. 2023). Through careful use of injection and recovery techniques, water treatment, monitoring technologies, and environmental stewardship, ASR significantly contributes to local water security, resilience, and sustainability amidst escalating water challenges (Page et al. 2011).

Water Conservation Techniques

Various techniques and innovations aim to conserve water resources (Fabbiano et al. 2020). Recent advancements include smart irrigation systems, leak detection technologies, and rainwater harvesting methods (Raimondi et al. 2023). Smart irrigation optimizes watering schedules based on soil moisture and weather forecasts using sensors and data analytics (Mohd Yussof and Ho 2022). Leak detection reduces waste by locating and fixing leaks in infrastructure and pipelines (Ertop et al. 2023; Rajpal et al. 2024). Rainwater harvesting captures and stores rainwater for later use, lessening reliance on conventional sources. Smart water grids combine automation and real-time data analytics to minimize losses and ensure efficient use (Tortajada 2020). Recycling and reusing water, facilitated by advanced treatment technologies, alleviate strain on freshwater sources. Desalination technologies are making freshwater production from salty water more energy-efficient and sustainable (Gray et al. 2011). Precision agriculture techniques like drip irrigation and soil monitoring optimize

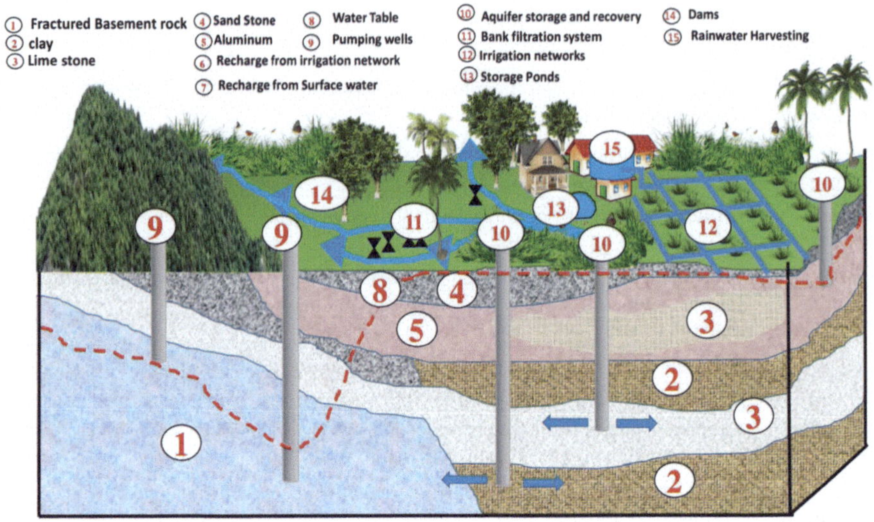

Fig. 8.2 Illustrates a model outlining various techniques for water management and conservation aimed at promoting sustainability

water use in farming (Bhardwaj et al. 2019). Mobile apps and internet platforms assist individuals and communities in tracking water consumption and adopting conservation habits, promoting sustainable water management practices (Fabbiano et al. 2020). Recent advancements in water conservation use smart irrigation systems, leak detection technologies, and rainwater harvesting methods (Bwambale et al. 2022). These technologies are designed to improve water management in agricultural, industrial, and domestic settings, ultimately leading to more efficient water usage and promoting sustainability and resilience (Pradhan et al. 2019). Smart irrigation systems optimize watering schedules and volumes based on variables like soil moisture levels and weather forecasts using sensors and data analytics (Pradhan et al. 2019; Raimondi et al. 2023). The use of sensors, controllers, and weather changes reduces water waste by delivering the right amount of water to crops or landscaping precisely when needed (Fig. 8.2) (Bwambale et al. 2022).

Wastewater Treatment and Recycling

Recent wastewater treatment and recycling technology developments contribute to water management (Gibson et al. 2018). Key technologies, including membrane filtration (Cevallos-Mendoza et al. 2022), sophisticated oxidation processes (Patel et al. 2020), and decentralised treatment systems enhance water quality and easing the repurposing of treated wastewater for non-potable uses, thereby mitigating the demand on freshwater resources (Armah et al. 2020; Saleem et al. 2024). Various techniques and innovations are aimed at conserving water resources (Table 8.2). Semi-permeable membranes are used in membrane filtration technologies to filter pollutants and impurities out of wastewater (Mansouri et al. 2010). By efficiently removing particles, bacteria, and other impurities from water, these membranes produce treated water that satisfies quality requirements and can be used for a variety of non-potable purposes (Le and Nunes 2016; Singha and Kumar Mishrab 2020; Cevallos-Mendoza et al. 2022). Advanced oxidation processes (AOPs) degrade and eliminate organic and inorganic contaminants from wastewater through chemical reactions using oxidising solid agents (Kiendrebeogo et al. 2021; Cai et al. 2022). These methods can effectively break down complex pollutants, producing treated wastewater with much lower pollutant levels (Altuhafi et al. 2021). By offering on-site or small-scale treatment options, decentralised treatment systems differ from conventional centralised wastewater treatment facilities (Tortajada 2020). Because of their scalability and adaptability, these systems can be used in various environments, such as industrial facilities, urban developments, and remote places (Bagatin et al. 2014). Community water quality can be improved, pollution can be decreased, and pressure on freshwater resources can be released using these cutting-edge wastewater treatments and recycling methods (Armah et al. 2020; Gogoi and Choudhury 2022; Ammar et al. 2023; Saleem et al. 2024). Reusing treated wastewater for non-potable uses like industrial processes, toilet flushing, and irrigation supports sustainable water management strategies and helps save precious freshwater resources for necessities (Chartzoulakis and Bertaki 2015).

Water Governance and Policy Innovations

To address the challenges in water resource management, this section redirects its attention towards the crucial role of policymakers in formulating strategies for sustainable water utilization (Brisbois and de Loë 2016; Pahl-Wostl et al. 2021). The discourse now centers on tactics for engaging stakeholders, incentive-driven pricing models, and comprehensive frameworks for overseeing water resources (Armitage et al. 2015). These strategies aim to ensure equitable distribution of water, promote collaboration, and enhance holistic decision-making (Brisbois and de Loë 2016). The comprehensive and coordinated approach for managing water across various sectors and stakeholders is termed IWRM (Armitage et al. 2015). Consideration is given to the interrelations among water-related challenges, such as the impact of urban development on water quantity and quality (Beveridge and Monsees 2012). Taking up an IWRM frame work aids policymakers in formulating strategies that preserve ecosystems and ensure sustainable water utilization while reconciling the conflicting demands of residential users, industry, and agriculture (Beveridge and Monsees 2012; Russo et al. 2014; Pahl-Wostl et al. 2021). Inclusive stakeholder involvement in decision-making processes is vital for effective water governance (Biswas

Table 8.2 Various water conversion techniques

Water conversion technique	Specified for water management
Desalination	Removes salt and impurities from seawater or brackish water to produce freshwater
Purification	Treats contaminated or polluted water to meet drinking water standards, typically using filtration and disinfection processes
Reverse osmosis	Uses a semi-permeable membrane to push water through it in order to eliminate impurities, salts, and pollutants
Distillation	Boils water to create steam, which is then condensed back into liquid form, leaving impurities behind
Ion exchange	Removes dissolved ions from water by exchanging them with similarly charged ions on an exchange medium
Electrocoagulation	Uses an electrical charge to destabilize suspended particles and contaminants, allowing for easier removal
UV disinfection	Exposes water to ultraviolet (UV) light to deactivate pathogens and microorganisms, making it safe for consumption
Ozonation	Treats water with ozone gas to oxidize and eliminate organic compounds, odors, and microorganisms
Membrane filtration	Utilizes membranes with specific pore sizes to physically separate impurities from water
Chemical treatment	Applies chemical agents such as chlorine, chlorine dioxide, or ozone to disinfect water and remove contaminants

2004). Stakeholder engagement methods promote collaboration among diverse entities, including local communities, businesses, governmental bodies, and environmental organizations (Beveridge and Monsees 2012). Inclusive decision-making leads to more sustainable and socially acceptable water management practices by ensuring that the viewpoints and concerns of multiple stakeholders are taken into consideration (Pahl-Wostl et al. 2021). Central to adopting incentive-based pricing strategies is adjusting water tariffs to encourage conservation and prudent usage (Beveridge and Monsees 2012). Tiered pricing systems can incentivize conservation by charging higher rates for higher consumption levels (Brisbois and de Loë 2016). Financial incentives or subsidies can also be introduced to encourage the adoption of water-efficient practices and technologies (Armitage et al. 2015). By implementing these innovative governance and policy approaches, communities can work towards more resilient and sustainable water management (Biswas 2004). In order to handle the complex issues surrounding water, these methods seek to improve cooperation, streamline decision-making procedures, and guarantee the fair and effective distribution of water resources (Fig. 8.3).

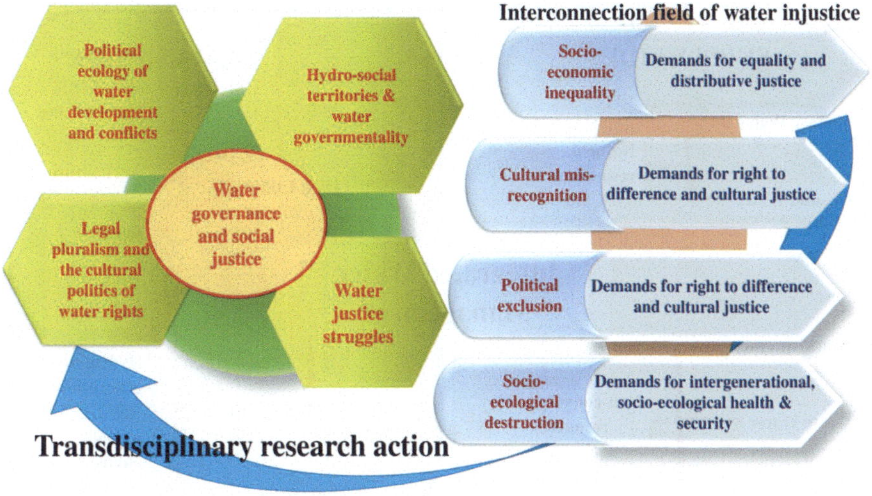

Fig. 8.3 Water governance and social justice and intervention

Integration of Technology in Water Resource Management

This section examines the use of sensing and computing technology in the management of water resources, emphasising the combination of artificial intelligence (AI), remote sensing, and data analytics (Geissen et al. 2015). With the help of these technical tools, stakeholders may successfully adapt to changing water concerns by monitoring water quality and quantity in real-time, modelling water availability ahead of time, and making well-informed decisions (Brisbois and de Loë 2016). Large amounts of water-related data, including streamflow, climate, and water quality observations, are analysed using data analytics techniques (Brisbois and de Loë 2016). Stakeholders can learn important information about trends, patterns, and possible problems affecting water resources by processing and analysing this data (Hargrove and Heyman 2020). Furthermore, framework for integrated water management encompassing both supply and demand-side utilization (Fig. 8.4). Data analytics provides decision-makers, water managers, and other stakeholders with valuable insights to inform evidence-based decision-making (Hargrove and Heyman 2020). Remote sensing technology, including satellite imaging and aerial surveys, plays a crucial role in monitoring water resources from a distance (Li et al. 2023). These tools offer both temporal and spatial data on variables such as surface water dynamics, vegetation health, and land cover (Chowdhury et al. 2023). Remote sensing data enables the evaluation of fluctuations in water availability, identification of areas susceptible to contamination or water scarcity, and monitoring the impacts of land use changes on water resources (Yang et al. 2022). Moreover, AI systems are increasingly utilized for tasks related to water resource management, such as flood forecasting, water quality prediction, and infrastructure optimization (Mahardhika and

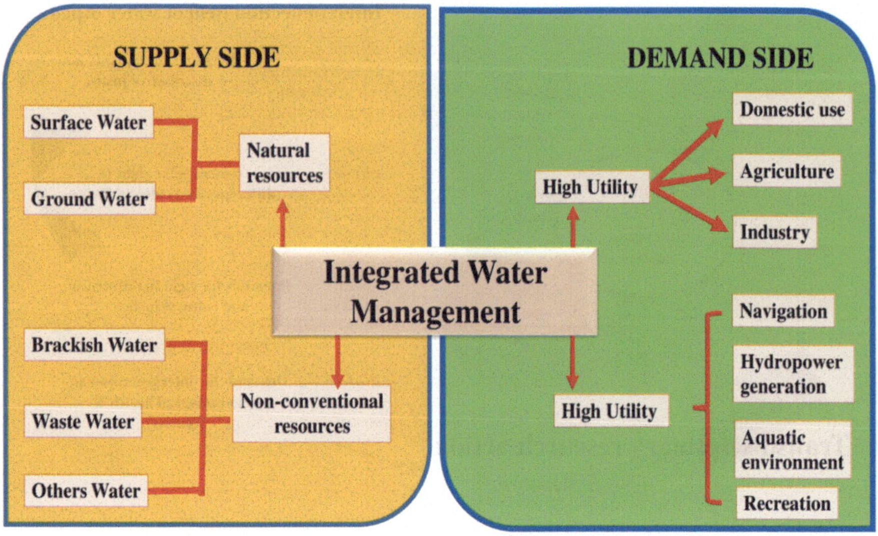

Fig. 8.4 Integrated water management for supply and demand-side utilization

Putriani 2023). Machine learning algorithms can be employed to develop predictive models for various water-related variables by analyzing historical data and identifying patterns and connections (Hulayyil et al. 2023). AI-driven decision support systems provide stakeholders with real-time insights and recommendations derived from data analysis, facilitating informed decision-making (Yang et al. 2022; Mahardhika and Putriani 2023). By leveraging these technologies, stakeholders can enhance their ability to monitor, manage, and sustainably utilize water resources (Yang et al. 2022). The long-term sustainability of water supplies is ensured by the combination of data analytics, remote sensing, and artificial intelligence (AI), which empowers stakeholders to confront new issues and make well-informed decisions (Mahardhika and Putriani 2023).

Future Directions and Recommendations

Innovative strategies are urgently needed to improve water resource management and confront water scarcity, pollution, and unequal access. Through the proactive utilization of technological advancements, pioneering policies, and collaborative frameworks, communities can forge resilient water systems that cater to both present needs and future demands. Various strategies, such as ASR, water conservation methods, wastewater treatment and recycling, progressive water governance, and technology integration, can support sustainable water management practices. By prioritizing these areas, societies worldwide can chart a course towards a more sustainable and equitable water future, safeguarding this precious resource for future generations.

Conclusion

In summary, this analysis underscores the critical imperative of embracing innovative strategies in water resource management to confront the multifaceted challenges of water scarcity, pollution, and unequal access. Through the proactive utilization of technological advancements, pioneering policies, and collaborative frameworks, communities can forge resilient water systems that cater to both present needs and future demands. The identified focal points, such as aquifer storage and recharge, water conservation techniques, wastewater treatment and recycling, progressive water governance, and technological integration, serve as pillars upon which sustainable water management practices can be built. By prioritizing these areas, societies worldwide can chart a course towards a more sustainable and equitable water future, safeguarding this precious resource for future generations.

References

Abdelfattah I, El-Shamy AM (2024) Review on the escalating imperative of zero liquid discharge (ZLD) technology for sustainable water management and environmental resilience. J Environ Manage 351:119614

Al-Hazmi HE, Mohammadi A, Hejna A et al (2023) Wastewater reuse in agriculture: prospects and challenges. Environ Res 236:116711

Alley WM (2009) Ground Water. In: Likens GE (ed) Encyclopedia of inland waters. Academic Press, Oxford, pp 684–690

Altuhafi A, Altun M, Hadwan MH (2021) The correlation between selenium-dependent glutathione peroxidase activity and oxidant/antioxidant balance in sera of diabetic patients with nephropathy. Rep Biochem Mol Biol 10:164–172

Amini A, Homayounfar V (2017) The groundwater balance in alluvial plain aquifer at Dehgolan, Kurdistan, Iran. Appl Water Sci 7:3113–3123

Ammar M, Yousef E, Mahmoud MA et al (2023) A comprehensive review of the developments in electrocoagulation for the removal of contaminants from wastewater. Separations 10:337

Armah EK, Chetty M, Adedeji JA et al (2020) Emerging trends in wastewater treatment technologies: the current perspective. In: Promising techniques for wastewater treatment and water quality assessment. IntechOpen

Armitage D, de Loë RC, Morris M et al (2015) Science–policy processes for transboundary water governance. Ambio 44:353–366

Bagatin R, Klemeš JJ, Reverberi AP, Huisingh D (2014) Conservation and improvements in water resource management: a global challenge. J Clean Prod 77:1–9

Bandyopadhyay S, Sharma A, Sahoo S et al (2021) Potential for aquifer storage and recovery (ASR) in South Bihar, India. Sustainability 13:3502

Bergström A, Hoa DP, Nga NT et al (2023) A facilitated social innovation: stakeholder groups using plan-do-study-act cycles for perinatal health across levels of the health system in Cao Bang Province, Vietnam. Implement Sci Commun 4:24

Beveridge R, Monsees J (2012) Bridging parallel discourses of integrated water resources management (IWRM): institutional and political challenges in developing and developed countries. Water Int 37:727–743
Bhardwaj A, Rajput R, Misra K (2019) Chapter 9—Status of arsenic remediation in India. In: Ahuja S (ed) Advances in water purification techniques. Elsevier, pp 219–258
Bhatt AG, Kumar A, Singh SK (2022) Hydro-geochemical evolution of groundwater and associated human health risk in River Sone subbasin of Middle-Gangetic floodplain, Bihar, India. Arab J Geosci 15:405
Biswas AK (2004) Integrated water resources management: a reassessment: a water forum contribution. Water Int 29:248–256
Borzì I (2023) Vulnerability assessment of water supply infrastructures through multiple indicator methodology. J Water Clim Change 14:3967–3984
Brisbois MC, de Loë RC (2016) State roles and motivations in collaborative approaches to water governance: a power theory-based analysis. Geoforum 74:202–212
Bwambale E, Abagale FK, Anornu GK (2022) Smart irrigation monitoring and control strategies for improving water use efficiency in precision agriculture: a review. Agric Water Manag 260:107324
Cai X, Zhang Z, Yin N et al (2022) Controlling microbial arsenite oxidation and mobilization in arsenite-adsorbed iron minerals: the Influence of pH conditions and mineralogical composition. J Hazard Mater 433:128778
Casanova J, Devau N, Pettenati M (2016) Managed aquifer recharge: an overview of issues and options. In: Jakeman AJ, Barreteau O, Hunt RJ et al (eds) Integrated groundwater management: concepts, approaches and challenges. Springer International Publishing, Cham, pp 413–434
Cevallos-Mendoza J, Amorim CG, Rodríguez-Díaz JM, Montenegro M da CBSM (2022) Removal of contaminants from water by membrane filtration: a review. Membranes (Basel) 12:570
Chartzoulakis K, Bertaki M (2015) Sustainable water management in agriculture under climate change. Agric Agric Sci Procedia 4:88–98
Chowdhury NR, Joardar M, Das A et al (2023) Appraisal of acute and chronic arsenic exposure in differently exposed school children with special reference to micronuclei formation in urine epithelial cells: a comparative study in West Bengal, India. Groundwater Sustain Dev 21:100917
Cosgrove WJ, Loucks DP (2015) Water management: current and future challenges and research directions. Water Resour Res 51:4823–4839
Deng C, Bailey RT (2022) Assessing the impact of artificial recharge ponds on hydrological fluxes in an irrigated stream-aquifer system. Hydrology 9:91
Duchenne-Moutien RA, Neetoo H (2021) Climate change and emerging food safety issues: a review. J Food Prot 84:1884–1897
Ertop H, Kocięcka J, Atilgan A et al (2023) The importance of rainwater harvesting and its usage possibilities: Antalya Example (Turkey). Water 15:2194
Fabbiano L, Vacca G, Dinardo G (2020) Smart water grid: a smart methodology to detect leaks in water distribution networks. Measurement 151:107260
Geissen V, Mol H, Klumpp E et al (2015) Emerging pollutants in the environment: a challenge for water resource management. Int Soil Water Conserv Res 3:57–65
Gibson MT, Campana ME, Nazy D (2018) Estimating aquifer storage and recovery (ASR) regional and local suitability: a Case Study in Washington State, USA. Hydrology 5:7
Gogoi N, Choudhury M (2022) Green materials synthesis for wastewater treatment. In: Encyclopedia of green materials. Springer Nature, Singapore, pp 1–14. https://doi.org/10.1007/978-981-16-4921-9_31-1
Gray S, Semiat R, Duke M et al (2011) 4.04—Seawater use and desalination technology. In: Wilderer P (ed) Treatise on water science. Elsevier, Oxford, pp 73–109
Hargrove WL, Heyman JM (2020) A comprehensive process for stakeholder identification and engagement in addressing wicked water resources problems. Land 9:119
Hauser-Davis RA, Wosnick N (2022) Climate change implications for metal and metalloid dynamics in aquatic ecosystems and its context within the decade of ocean sciences. Water 14:2415

Hulayyil SB, Li S, Xu L (2023) Machine-learning-based vulnerability detection and classification in internet of things device security. Electronics 12:3927

Inyinbor Adejumoke A, Adebesin Babatunde O, Oluyori Abimbola PO et al (2018) Water pollution: effects, prevention, and climatic impact. In: Water challenges of an urbanizing world. IntechOpen

Kiendrebeogo M, Karimi Estahbanati MR, Khosravanipour Mostafazadeh A et al (2021) Treatment of microplastics in water by anodic oxidation: a case study for polystyrene. Environ Pollut 269:116168

Kumar D, Choudhury M (2021) Recognizing economic values of wetland ecosystem services: a study of emerging role of monetary evaluation of Chandubi ecosystem and biodiversity. In: Wetlands conservation: current challenges and future strategies, pp 87–110. https://doi.org/10.1002/9781119692621.ch5

Le NL, Nunes SP (2016) Materials and membrane technologies for water and energy sustainability. Sustain Mater Technol 7:1–28

Li J, Zhao M, Han Y, Wei J (2023) Analysis of water resource ecological optimization and obstacles based on the water resources—socio-economic-ecological environment model framework—a case study of the core area of the Central Plains urban agglomeration. Front Ecol Evol 11

Liquete C, Udias A, Conte G et al (2016) Integrated valuation of a nature-based solution for water pollution control. Highlighting hidden benefits. Ecosyst Serv 22:392–401

Mahardhika SP, Putriani O (2023) Deployment and use of artificial intelligence (AI) in water resources and water management. IOP Conf Ser: Earth Environ Sci 1195:012056

Majumder R, Kaviraj A (2017) Cypermethrin induced stress and changes in growth of freshwater fish *Oreochromis niloticus*. Int Aquat Res 9:117–128

Maliva RG (2014) Economics of managed aquifer recharge. Water 6:1257–1279

Mansouri J, Harrisson S, Chen V (2010) Strategies for controlling biofouling in membrane filtration systems: challenges and opportunities. J Mater Chem 20:4567–4586

Martinho VJPD (2019) Best management practices from agricultural economics: mitigating air, soil and water pollution. Sci Total Environ 688:346–360

McGrane SJ (2016) Impacts of urbanisation on hydrological and water quality dynamics, and urban water management: a review. Hydrol Sci J 61:2295–2311

Mohd Yussof NA, Ho HW (2022) Review of water leak detection methods in smart building applications. Buildings 12:1535

Müller A, Österlund H, Marsalek J, Viklander M (2020) The pollution conveyed by urban runoff: a review of sources. Sci Total Environ 709:136125

Nestler A, Berglund M, Accoe F et al (2011) Isotopes for improved management of nitrate pollution in aqueous resources: review of surface water field studies. Environ Sci Pollut Res 18:519–533

Othman NH, Alias NH, Fuzil NS et al (2022) A review on the use of membrane technology systems in developing countries. Membranes 12:30

Page D, Miotliński K, Dillon P et al (2011) Water quality requirements for sustaining aquifer storage and recovery operations in a low permeability fractured rock aquifer. J Environ Manage 92:2410–2418

Pahl-Wostl C, Dombrowsky I, Mirumachi N (2021) Water governance and policies. In: Bogardi JJ, Gupta J, Nandalal KDW et al (eds) Handbook of water resources management: discourses, concepts and examples. Springer International Publishing, Cham, pp 253–272

Patel AB, Shaikh S, Jain KR et al (2020) Polycyclic aromatic hydrocarbons: sources, toxicity, and remediation approaches. Front Microbiol 11:562813

Pradhan R, Sahoo J, Pradhan R, Sahoo J (2019) smart rainwater management: new technologies and innovation. In: Smart urban development. IntechOpen

Prusty P, Farooq SH (2020) Seawater intrusion in the coastal aquifers of India—a review. HydroResearch 3:61–74

Raimondi A, Quinn R, Abhijith GR et al (2023) Rainwater harvesting and treatment: state of the art and perspectives. Water 15:1518

Rajpal A, Ali M, Choudhury M, Almohana AI, Alali AF, Munshi FMA, Khursheed A, Kazmi AA (2022) Abattoir wastewater treatment plants in India: understanding and performance evaluation. Front Environ Sci 10:881623. https://doi.org/10.3389/fenvs.2022.881623

Rajpal A, Choudhury M, Goswami S, Chakravorty A, Raghavan V (eds) (2024) Waste management and treatment: advances and innovations, 1st edn. CRC Press. https://doi.org/10.1201/9781003258377

Rauh E, Hughes S (2024) Collaboration for source water protection in the United States: community water systems engagement in nitrate pollution reduction. Wires Water 11:e1682

Russo T, Alfredo K, Fisher J (2014) Sustainable water management in urban, agricultural, and natural systems. Water 6:3934–3956

Saleem MH, Mfarrej MFB, Khan KA, Alharthy SA (2024) Emerging trends in wastewater treatment: addressing microorganic pollutants and environmental impacts. Sci Total Environ 913:169755

Samal AK, Mishra PK, Biswas A (2020) Assessment of origin and distribution of fluoride contamination in groundwater using an isotopic signature from a part of the Indo-Gangetic Plain (IGP), India. HydroResearch 3:75–84

Sasidharan S, Bradford SA, Šimůnek J, Kraemer SR (2021) Comparison of recharge from drywells and infiltration basins: a modeling study. J Hydrol 594:125720

Shawaqfah M, Almomani F, Al-Rousan T (2021) Potential use of treated wastewater as groundwater recharge using GIS techniques and modeling tools in Dhuleil-Halabat well-field/Jordan. Water 13:1581

Singha I, Kumar Mishrab P (2020) Nano-membrane filtration a novel application of nanotechnology for waste water treatment. Mater Today Proc 29:327–332

Tortajada C (2020) Contributions of recycled wastewater to clean water and sanitation sustainable development goals. NPJ Clean Water 3:1–6

Wang Q, Wang X, Liu Y, Li R (2021) Urbanization and water consumption at national- and subnational-scale: the roles of structural changes in economy, population, and resources. Sustain Cities Soc 75:103272

Yang H, Kong J, Hu H et al (2022) A review of remote sensing for water quality retrieval: progress and challenges. Remote Sens 14:1770

Zavala-Yoe R, Iqbal HMN, Ramirez-Mendoza RA (2020) Understanding the evolution of pollutants via hierarchical complexity of space-time deterministic and stochastic dynamical systems. Sci Total Environ 710:136245

Part III
Strengthening Ecosystems: Policy Frameworks and Governance for Resilience

Chapter 9
Climate Resilience from the Ground Up: Community and Indigenous Insights—A Case Study from Developing Nation

Pankaj Kumar, Godfred Addai, Mukul Machhindra Barwant, Lawrence Guodaar, and Ajar Nath Yadav

Abstract Indigenous groups emerge as essential guardians of a priceless natural heritage. These groups have developed good insights from generations coexisting with nature firmly rooted in their natural environment. Their commitment to nature preservation creates a complex network of understanding and practices that protect their local environment, flora, and wildlife. Also, Indigenous groups have insights into climate change adaptation by exhibiting several practices that help them to understand the variations in average temperature, wind, solar radiation, rainfall intensity, and patterns. This is pertinent because local communities and Indigenous people are experiencing real, continuing effects of climate change that are affecting many different aspects of their socio-ecological systems. Since agriculture has been the bulk of the population's primary source of income, these effects tend to render the Indigenous population and the community vulnerable. Notwithstanding challenges and information shortages, these approaches may be strengthened by accepting and modifying newly imported technology from other societies or

P. Kumar (✉)
Department of Veterinary Physiology, College of Veterinary and Animal Science, RAJUVAS, Bikaner, Bikaner, Rajasthan 334001, India
e-mail: pankajmukundgarh@gmail.com

G. Addai
Department of Planning, Planning Institute of Australia, Barton, Australia

M. M. Barwant
Department of Botany, Sanjivani Arts Commerce and Science College, Kopargaon, Maharashtra, India

L. Guodaar
Department of Geography and Rural Development, Kwame Nkrumah University of Science and Technology, Kumasi, Ghana

A. N. Yadav
Department of Genetics, Plant Breeding and Biotechnology, Dr. Khem Singh Gill Akal College of Agriculture, Eternal University, Baru Sahoib, Sirmour, Himachal Pradesh 173101, India

M. Choudhury et al. (eds.), *Sustainable Synergy: Harnessing Ecosystems for Climate Resilience*, Climate Change Management,
https://doi.org/10.1007/978-3-031-77957-2_9

contemporary science. Therefore, it is necessary to identify these behaviours since community-based adaptation strategies could successfully mainstream them.

Keywords Indigenous groups · Climate change · Livestock · Conservation · Water harvesting

Introduction

An estimated 300–500 million indigenous peoples, or 5% of the world's population, live in about 90 different countries (Chao 2012). The traditional indigenous territories include tropical forests, small islands, high-altitude regions, beaches, the margins of deserts, and the circumpolar Arctic, making up 22% of the planet's total geographical area. Approximately 25% of all species on Earth are found in these territories (World Resources 2005). As a result, most biodiversity hotspots worldwide are found in areas where Indigenous people live. This demonstrates unequivocally that Indigenous people's traditional knowledge, wisdom, and behaviours have allowed them to preserve their biodiversity (Martemjen 2017).

Despite having made minimal contributions to the primary causes of climate change, indigenous peoples suffer the most from its repercussions. Their long-standing reliance on the region's biological diversity, natural resources, and historic sites for survival, resilience, and well-being is the primary reason for this. Indigenous peoples' identity is deeply linked to their original home. Climate change has an impact on traditional activities such as pastoralism, agriculture, hunting, fishing, and gathering, which rely on water availability in areas where the dominant civilization is encroaching on the world's biocultural heritage. Fortunately, Indigenous peoples are exceptional observers and interpreters of environmental change due to their unique collective knowledge of the land, sky, and sea (Egeru 2012; Kumar 2014). Figure 9.1 shows traditional activities affected by climate change.

However, because this information is insufficiently documented and treated with disrespect, it is rarely used when planning and executing modern mitigation and adaptation plans. Indigenous knowledge systems have existed for millennia without being documented. Rebuilding and sustaining indigenous knowledge networks is crucial for solving current concerns like climate change. More and more people recognize the value of preserving indigenous knowledge in order to sustain livelihoods and food security, especially in the face of unexpected or changing weather patterns.

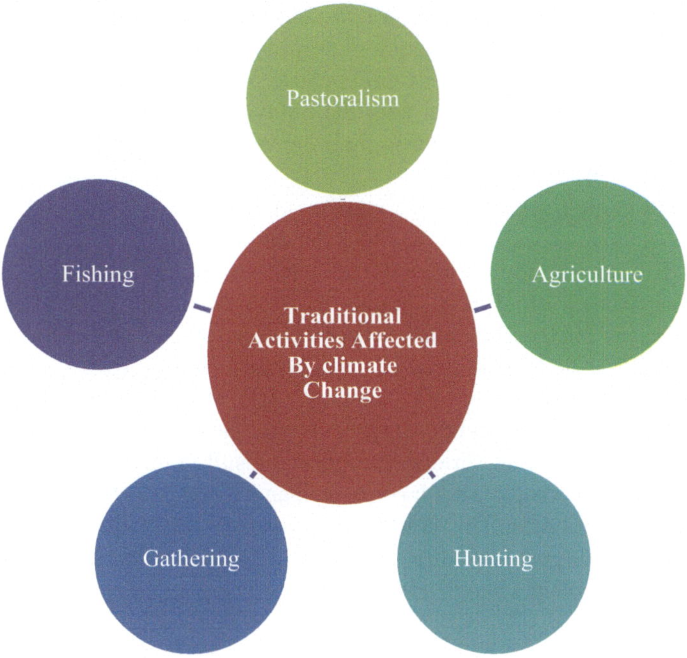

Fig. 9.1 Traditional activities affected by climate change

Climate Change Adaptation by Indigenous Communities

The Environmental Change Institute in Oxford, United Kingdom, brought together researchers in 2007 from a wide range of disciplinary backgrounds, including the humanities, social sciences, physical and natural sciences, academic and non-profit organizations, to discuss how local people, particularly indigenous people, are impacted by climate change and how they respond to and interpret these changes. Indigenous peoples' suffering was depicted, but so was their inventiveness and constructive responses to climate change (Salick and Ross 2018).

By altering the climate, communities actively adjust to the new conditions rather than merely being passive observers (Salick and Byg 2007; Boedhihartono et al. 2008). Indigenous people are finding ways to adapt to climate change and cope with the loss of biodiversity through a variety of strategies, such as irrigation, migration, conservation of water, reclamation of land, changing the when, where, and elevation at which plants are cultivated, and adaptation in order to survive (Boedhihartono et al. 2008). Another example would be the different approaches to common floods that communities with access to traditional ecological knowledge have developed to deal with the numerous effects of climate change, which are often caused by irregular and unpredictable patterns of rainfall. Prevention or adaptation techniques including terracing, riverbank retaining walls, biostabilizing slopes using native plants and rocks, and other techniques for flood resistance have been the result of years of rigorous research. These safety precautions are also put in place to lessen the harm that mudflows, rock falls, and landslides might cause. Indigenous people have "evolved fine-tuned social systems" to deal with the unfavorable adverse effects of natural disasters. These systems are based on their conventions and traditional ecological knowledge (Byg and Salick 2009).

Traditional ecological knowledge can be applied to environmental and social impact assessments, as well as to the understanding and interpretation of ecological processes. To find out how Indigenous people's traditional ecological knowledge might further studies on climate change, more investigation is necessary (Fikret 1999; Huntington 2000). The inhabitants of Pacific atolls are able to recognize and respond to small-scale local climate changes because of their traditional ecological knowledge system (Bridges and McClatchey 2009). Models such as this one could be used to develop more thorough responses to climate change. Himalayan peasants are now cultivating a greater variety of crops as a result of their attempts to cultivate hitherto uncultivated crops. This has also expanded the selection of native foods available at high altitudes. This could aid in people's adaptation to the effects of climate change having an impact on their way of life (Chaudhary and Bawa 2011). Figure 9.2 shows climate change adaptations.

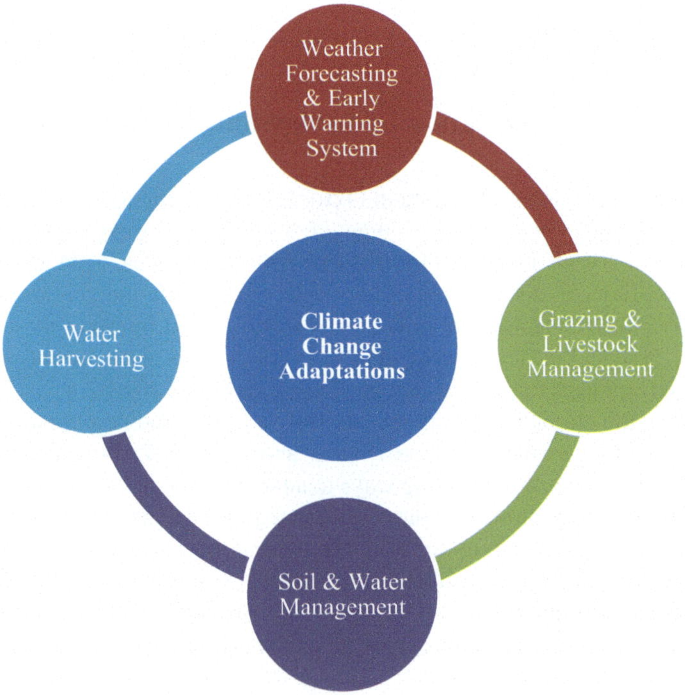

Fig. 9.2 Climate change adaptations

Indigenous knowledge is crucial in helping local communities understand environmental changes, particularly in microclimates, which occur locally. As a result of environmental change, more communities are using traditional weather forecasting methods to adjust to changing weather patterns. These methods have mainly been used to forecast seasonal climate change, such as when rain will start, where it will fall most heavily, and when it will stop. Because these climatic patterns frequently have a "finer" spatiotemporal resolution than current climate forecasts, pastoralists have found significance in these patterns (Luseno et al. 2003). Because local forecasting methods are based on personal experience, farmers and communities rely on them. Therefore, communities can adjust to a changing environment thanks to their indigenous forecasting techniques.

Weather Forecasting and Early Warning System

Droughts and floods are examples of natural calamities that can seriously affect communities. Hydro-meteorological disasters pose a particular threat to Southeast Asian archipelagos and small island nations (Hiwasaki et al. 2015). Various civilizations employ various methods and indicators to forecast the weather. Until modern scientific methods were applied for weather forecasting and climate prediction, farming and other elements of daily life were supported by traditional wisdom to deal with the unpredictability of the climate, the traditional inhabitants of the area still mostly rely on their indigenous knowledge systems. In Uttarakhand, India, the valleys of Johar, Byans, Bhagirathi, and Niti use bioindicators to anticipate the weather in addition to physical parameters like wind direction, time of year, sky color, and star position (Rautela and Karki 2015). To predict the weather for the upcoming days, they closely monitor the changes in their natural surroundings. In Rajasthan, India, the arrival of butterflies marks the start of a favorable season and the early arrival of rain.

Grazing and Livestock Management

Pastoral groups are better able to make educated judgments about land use due to their "in-depth knowledge" of indigenous rangeland assessment procedures. Timely grazing of their "multi-livestock species" is decided by herders through daily inspections and monitoring of rangelands' conditions (Homewood and Rodgers 1991; Cotton 1996; Mills et al. 2002). The Kumaon Himalayan's Bhotiya pastoralists, located in central India, have distinguished themselves from other Kumaonites by their farming and livestock management practices throughout several generations. Bhotiya have been grazing their cattle through migration and vertical transhumance for ages (Farooquee and Rao 1999). Pastoralists who live close to mountains use a crucial animal husbandry technique called vertical transhumance, which helps to ensure that animals have a consistent supply of high-quality pasture all year round (Makarewicz et al. 2017). From May to September, when it's cooler, professional shepherds move the Bhotiya sheep and goats to the foothills and plains. Legal limits allow only shepherds to utilize a pasture for ten to fifteen days at a time (Nusser and Gerwin 2008). For grazing, in China the most popular areas are the grasslands of the Inner Mongolian Plateau (Liu 2017). Before the decentralisation of herders and the privatisation of grasslands, pastoralists in Inner Mongolia had a nomadic lifestyle that focused on seasonal migrations, which enabled grasslands to recuperate from use (Conte 2015).

The Mongolian people's traditional ecological knowledge, which takes into account the best plant species for cattle as well as seasonal food and water requirements, informs the decision of when and where to move the herd. This nomadic livestock management approach greatly aids the sustainable management of the grazing land in Inner Mongolia. It encourages the dispersion of herds to suitable pastures and the effective use of grassland resources (Conte 2015).

Management of Soil and Water

The preservation of soil and water resources is the main objective of sustainable development (Gogoi et al. 2024a, b). Topsoil, which is an important resource for crops because it contains essential nutrients and organic matter, can be significantly lost as a result of soil erosion. Farmland loses its ability to hold water as a result of losing organic matter and soil nutrients, which makes the land less suited for cultivation. Long-term resource users among indigenous people are typically quite aware of how complex natural systems behave in their surroundings (Gogoi et al. 2024a, b).

In order to maintain soil moisture, farmers also employ a variety of ploughing methods. This can facilitate ideal seed germination by preserving soil moisture in the soil's lower layers. Fields are ploughed soon after the rainy season's crop harvest. In order to avoid competition for water while preparing the seedbed, farmers also saturate fallow fields with water in October. In the area, paddy nurseries are also cultivated in a circular fashion through seed sowing. This technique is known for its capacity to save water when compared to line sowing and broadcasting (Negi and Kandpal 2003).

Farmers in the Himalayas employ floodwater because it deposits fine silts that improve soil fertility and carry humus and organic matter (Sharma and Choudhury 2021). In the field of crops, they also engage in in-situ maturation. Days following the Kharif crop harvest, the practice entails leaving cattle (or other livestock) out all night in the field. The animals are relocated to different fields when the amount of faeces and urine in the fields is deemed sufficient. These farmers also employ moist soils combined with rye (*Brassica nigra*) seeds between the spaces between terrace riser stones to reduce the amount of water required for germination (Negi and Kandpal

2003). Biophysical treatments are not the only tactics that may be used to maintain soil fertility. The elder generation typically decides to stay and uses inorganic fertiliser in combination with either animal manure or domestic trash, or both, the older generation chooses to remain and uses inorganic fertiliser in combination with either animal manure, domestic trash, or both to improve soil production (Saidou 2006).

Techniques for Harvesting and Preserving Water

In practice and function, fifteen key categories of ancient and traditional Rainwater Harvesting were similar, even if they were found in different climatic zones or among different cultures (Table 9.1) (Akpinar Ferrand and Cecunjanin 2014). Figure 9.3 shows rainwater harvesting techniques (Figs. 9.4, 9.5 and 9.6).

Barriers to Climate Change Adaptation Among the Indigenous Communities

As the community's elders, who serve as the primary keepers of this knowledge, are dwindling in number, if indigenous knowledge is not sufficiently documented, it is currently in danger of being extinct. Regretfully, younger generations are starting to see Indigenous knowledge as outdated and useless, with little possibility of being included in the development strategies now in place. Figure 9.7 shows barriers to climate change.

Youths Attitude Towards Technologies and Practices

Indigenous knowledge and traditional ecological knowledge are marginalised for a variety of reasons (Ocholla 2007).

- Because the knowledge is not consistently recorded or codified, it is challenging to transmit or transfer,
- The knowledge is oral mainly and only exists in the beholder's memory, if it is not transmitted, it perishes with the beholder,
- The information is not universal and is therefore challenging to globalise since it is ingrained in the customs, beliefs, language, and religion of a specific particular society,
- A multicultural, metropolitan, and economically well-off community would not benefit from the knowledge, which is primarily rural and typically practised among impoverished communities,

Table 9.1 Rainwater harvesting techniques

Rainwater harvesting techniques	Detail
The formation of pits	On level or sloped ground, cultivars are planted in excavated pits. Pitting helps to improve groundwater recharge and preserve soil moisture. Pitting will slow down precipitation-induced runoff if it is built along a contour
Creating contours using vegetative barriers, hedgerows, and stone or soil bunds	Contouring is the process of laying stone or earthen banks in a cultivated hillside according to a contour. To create hedgerows, grass and plants are typically planted in cross-slope barriers or along a contour. Three advantages of contouring include minimizing soil erosion and shortening the slope's length
Terracing	Bunds connected to a trench on a sloping surface along a contour. They assist in reducing erosion, preserving soil moisture, and lowering runoff velocity during precipitation events
Micro basins	Various shaped little basins are encircled by earthen or stone bunds to allow precipitation-related runoff to pass through
Pit courtyard	Because pit courtyards are encircled by walls and other impermeable surfaces, they serve as an impluvium to collect rainfall
In situ rainwater harvesting	By conserving tillage, mulching, broad-bed and furrowing, ridging, and hoeing, these practices seek to reduce soil evaporation and enhance precipitation infiltration
Rooftop rainwater harvesting	Rainwater is gathered from roofs and stored in storage tanks that are placed in backyards of homes
Common open pond	Runoff and precipitation are gathered in open ponds by a catchment system. The pond's linings might be seen or not. Seepage could be the result of a lining break
Cisterns	Runoff is gathered and kept in subterranean storage reservoirs. Old cisterns could be naturally occurring or artificially created pits in solid rock
Minor dams	A controlled flow of rainwater that has been stored and is meant to seep behind soil or stone barriers on a sloping ground. In order to benefit from silt deposition and elevated soil moisture after drainage, certain micro-dam surfaces are planted
Shallow water wells	In low depressions or ponds, shallow wells are dug to collect surface runoff after percolation in order to receive water during the dry season

(continued)

Table 9.1 (continued)

Rainwater harvesting techniques	Detail
System of underground wells	A local network of rain-fed horizontal channels, or galleries, that are carved into an alluvial fan aquifer at the base of a mountain
Diversion of runoff and spate irrigation	Seasonal flooding that begins on farms and spreads to isolated spots of precipitation. Dams of different sizes, reservoir systems, and terraces can all be connected to these systems
Building of dams	Around the foot slope of the hill, enormous storage systems are constructed to collect and manage runoff, rainwater, and ephemeral streams
Large lakes or reservoirs	Wetness and runoff collection in large man-made artificial basins

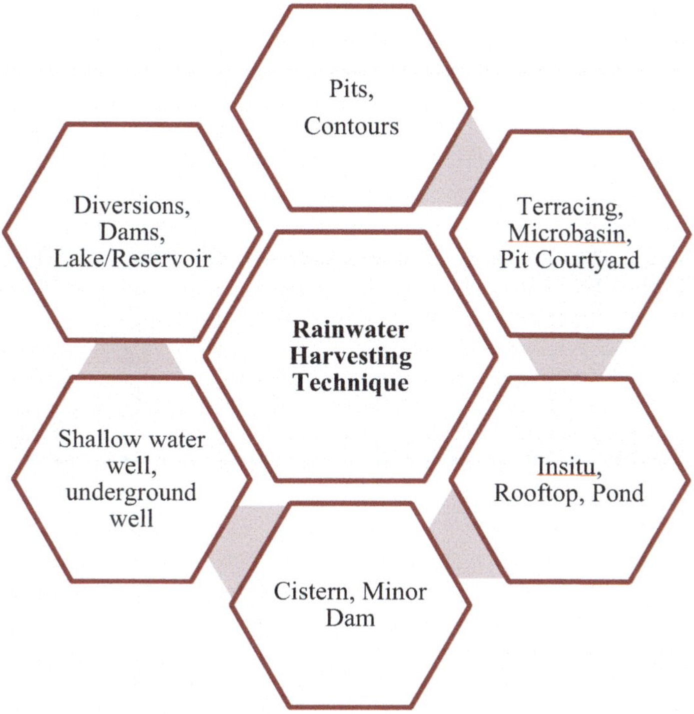

Fig. 9.3 Rainwater harvesting techniques

Fig. 9.4 Structure of a dam

Fig. 9.5 Large reservoir

- The information is viewed as outdated and archaic and as such, is of little use,
- Communities and families are growing more fragmented and globalised, and
- There is no scientific foundation for some of the arguments highlighted.

Challenges for Indigenous Knowledge System

Despite the effectiveness of Indigenous knowledge and its verifiable facts, certain challenges still need to be resolved (Fabiyi and Oloukoi 2013). The transferability of Indigenous knowledge from generation to generation and scientific efficiency" is one example that could come up. This facilitates the loss of these undocumented

Fig. 9.6 Large lake

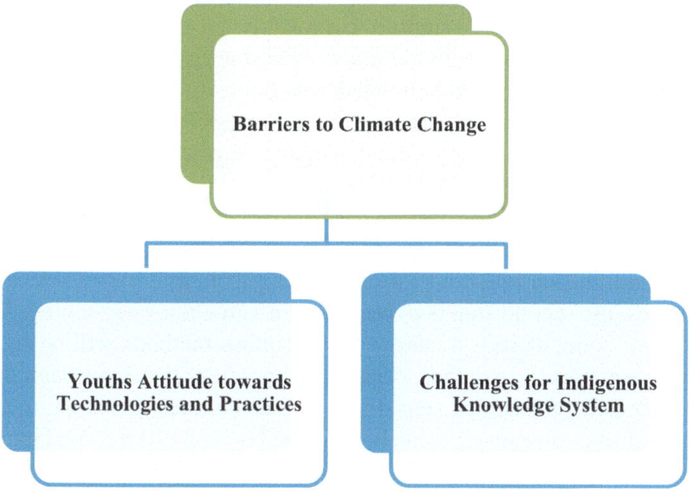

Fig. 9.7 Barriers to climate change

knowledge systems. Indigenous peoples have historically used age-grade initiations, community celebrations, and folktales to pass on their wisdom to new generations. But the children in the villages will never know this, because all these oral traditions are now seen by the younger generations as local and antiquated. For traditional civilizations, adjusting to the unpredictable nature of climate change is nothing new. Communities have benefited from both indigenous coping methods and state-of-the-art adaptation measures for the present and future climate unpredictability (Kumar and Choudhury 2021).

Policy Recommendations

The ultimate manifestation of humanity's growing alienation from the environment is anthropogenic climate change (Green and Raygorodetsky 2010). While not entirely to blame for the current situation are the Indigenous people who reside on their ancient grounds, and anticipated effects of climate change, they are nonetheless facing a wide range of implications on their resources and territory. Due to their increased reliance on the natural environment and decreased socioecological resilience, they undoubtedly endure both direct and indirect effects of climate change. Indigenous peoples face a multitude of social, environmental, political, and economic challenges, making them especially susceptible to the consequences of climate change. These challenges endanger their way of life, communities, and intimate ties to the natural world (Maldonado et al. 2016).

Indigenous people can offer a variety of techniques and methods to address current climate change since they have historically been able to adapt to the climatic unpredictability in their area. As a result, it's important to increase Indigenous knowledge and use modern techniques that help them adapt to and fight climate change. Rather than encouraging more scientific conformity. Decision-making demands the mobilisation of the greatest available knowledge due to the scope and complexity of today's situations. Indigenous knowledge is contextual and local, passed down orally, by imitation and example, adaptable to changing conditions, collected through a common social memory, and embedded in many interconnected aspects of peoples' lives (Mistry and Berardi 2016; Butler 2019).

The first step in addressing climate change adaptation in agriculture is to catalog the indigenous practices and customs of rural communities as a coping technique. In order to highlight the potential for exchanging techniques across related agroecological zones, the second step is to establish an agroecology grouping. Within the same biological zone, there is a chance that adaption methods will be successfully adapted from one place to another. Adaptation practices that have been discovered will be indigenized in the third step. Improved techniques for adapting to climate change that address water scarcity can be informed by newly developed interventions based on locally well-adapted naturally sustainable Indigenous practices (Denison and Wotshela 2012). Indigenous or traditional knowledge should help recognise the possibilities of some affordable, participatory, and sustainable adaptation techniques (Intergovernmental Panel on Climate Change (IPCC) 2018).

Conclusion

Climate change is one of the main risks to life as we know it. The main ways it manifests itself are through variations in the mean temperature, wind direction and speed, solar radiation, and the amount and pattern of rainfall. These changes significantly impact essential resources like water and land resources. Dependent communities in the agricultural or coastal sectors, as well as marginalized and vulnerable groups like Indigenous peoples, are disproportionately more susceptible to adverse consequences from global warming of 1.5 °C and higher (Intergovernmental Panel on Climate Change (IPCC) 2018). Consequently, it is recommended that adaptation measures be taken to address climate change. Because they have personally experienced the detrimental effects of climate change, indigenous peoples have created strategies for adapting to it. It was essential to identify community-based adaption measures since they could successfully mainstream certain behaviors. Weather forecasting, early warning systems, water harvesting, and storage techniques were technical and practice themes connected to each other's agroecological zones (AEZ). The AEZ strategy was considered a valuable place to start when adopting or amending an established indigenous plan for similar locations. For Indigenous and community adaptation strategies to be successful, a few challenges must be solved. These challenges include how climate change affects community resources and indicators, how the population growth and influx of animals are straining natural resources beyond their capacity to withstand shocks, how little information is being shared with younger communities, and so forth. Indigenous knowledge is indispensable for community-based adaptation measures to maintain the resilience of social-ecological systems at the interrelated local, regional, and global levels. These tactics can be reinforced by accepting and adapting recently brought technologies from foreign cultures or modern science despite obstacles and gaps in knowledge. Several organizations that receive sponsorships know about these tactics, even if they are not formally organized.

References

Akpinar Ferrand E, Cecunjanin F (2014) Potential of rainwater harvesting in a thirsty world: a survey of ancient and traditional rainwater harvesting applications. Geogr Compass 8(6):395–413

Boedhihartono A, Wolfangel C, Howell M (2008) Indigenous and traditional peoples and climate change

Bridges KW, McClatchey WC (2009) Living on the margin: ethnoecological insights from Marshall Islanders at Rongelap Atoll. Glob Environ Chang 19(2):140–146

Butler G (2019) United Nations Educational, Scientific and Cultural Organization (UNESCO). In: Research handbook on the European Union and international organizations. Edward Elgar Publishing, pp 142–164

Byg A, Salick J (2009) Local perspectives on a global phenomenon—climate change in Eastern Tibetan villages. Glob Environ Chang 19(2):156–166

Chao S (2012) Forest peoples: numbers across the world, vol 10. Forest Peoples Programme, Moreton-in-Marsh, UK, p 2017

Chaudhary P, Bawa KS (2011) Local perceptions of climate change validated by scientific evidence in the Himalayas. Biol Let 7(5):767–770

Conte TJ (2015) The effects of China's grassland contract policy on Mongolian herders' attitudes towards grassland management in northeastern Inner Mongolia. J Polit Ecol 22(1):79–97

Cotton CM (1996) Ethnobotany: principles and applications. Wiley

Denison JA, Wotshela L (2012) An overview of indigenous, indigenised and contemporary water harvesting and conservation practices in South Africa. Irrig Drain 61:7–23

Egeru A (2012) Role of indigenous knowledge in climate change adaptation: a case study of the Teso Sub-Region, Eastern Uganda

Fabiyi OO, Oloukoi J (2013) Indigenous knowledge system and local adaptation strategies to flooding in coastal rural communities of Nigeria. J Indigenous Soc Dev 2(1)

Farooquee NA, Rao KS (1999) Conservation and utilization of indigenous cattle and livestock among the transhumant pastoralists of Kumaun Himalaya (India). J Environ Syst 27(4)

Fikret B (1999) Sacred ecology: traditional ecological knowledge and resource management. Taylor & Francis, London & Philadelphia

Gogoi N, Choudhury M, Ul Hasan MS, Changmai B, Baruah D, Samanta P (2024a). Soil organic carbon pool in diverse land utilization patterns in North-East India: an implication for carbon sequestration. Environ Dev Sustain, 1–27. https://doi.org/10.1007/s10668-024-05037-y

Gogoi N, Sarma A, Choudhury M, Samanta P, Faizan M, Sadhak S (2024b) Soil heavy metal pollution and ecological risk assessment in disturbed and undisturbed soil of Morigaon, Assam. Discover Environ 2(1):13. https://doi.org/10.1007/s44274-024-00039-z

Green D, Raygorodetsky G (2010) Indigenous knowledge of a changing climate. Clim Change 100(2):239

Hiwasaki L, Luna E, Syamsidik, Marçal J A (2015) Local and indigenous knowledge on climate-related hazards of coastal and small island communities in Southeast Asia. Clim Change 128:35–56

Homewood KM, Rodgers WA (1991) Maasailand Ecology: pastoralist development and wildlife conservation in Ngorongoro. Cambridge University Press, Tanzania

Huntington HP (2000) Using traditional ecological knowledge in science: methods and applications. Ecol Appl 10(5):1270–1274

Intergovernmental Panel on Climate Change (IPCC) 2018 Summary for Policymakers—Global warming of 1.5 °C. (n.d.) C. https://www.ipcc.ch/sr15/chapter/spm/

Kumar D, Choudhury M (2021) Recognizing economic values of wetland ecosystem services: a study of emerging role of monetary evaluation of Chandubi ecosystem and biodiversity. In: Wetlands conservation: current challenges and future strategies, pp 87–110. https://doi.org/10.1002/9781119692621.ch5

Kumar V (2014) Role of indigenous knowledge in climate change adaptation strategies: a study with special reference to North-Western India. J Geogr Nat Disast 5:131

Liu M (2017) China's grassland policies and the Inner Mongolian grassland system. Doctoral dissertation, Wageningen University and Research

Luseno WK, McPeak JG, Barrett CB, Little PD, Gebru G (2003) Assessing the value of climate forecast information for pastoralists: evidence from Southern Ethiopia and Northern Kenya. World Dev 31(9):1477–1494

Makarewicz CA, Arbuckle BS, Oztan A (2017) Vertical transhumance of sheep and goats identified by intra-tooth sequential carbon (δ13C) and oxygen (δ18O) isotopic analyses: evidence from Chalcolithic Kosk Hoyuk, central Turkey. J Archaeol Sci 86:68–80

Maldonado J, Bennett T B, Chief K, Cochran P, Cozzetto K, Gough B, Voggesser G (2016) Engagement with indigenous peoples and honoring traditional knowledge systems. In: The US National Climate Assessment: innovations in science and engagement, pp 111–126

Martemjen J (2017) Biodiversity conservation, indigenous knowledge and practices: a Naga perspective. Notion Press

Mills D, Blech R, Gillam B, Martin M, Fithardinge G, Davies J, Woodhams L (2002) Rangelands: people, perceptions and perspectives. In: Global rangelands: progress and prospects. CAB International, Wallingford, Oxfordshire, UK, pp 43–54

Mistry J, Berardi A (2016) Bridging indigenous and scientific knowledge. Science 352(6291):1274–1275

Negi GCS, Kandpal KD (2003) Traditional methods of water management in the Central Himalayan agriculture

Nusser M, Gerwin M (2008) Diversity, complexity and dynamics: land use patterns in the Central Himalayas of Kumaon, Northern India. In: Diversity in mountain systems, pp 107–119

Ocholla D (2007) Marginalized knowledge: an agenda for indigenous knowledge development and integration with other forms of knowledge. Int Rev Inf Ethics 7:236–245

Rautela P, Karki B (2015) Weather forecasting: traditional knowledge of the people of Uttarakhand Himalaya. J Geogr Environ Earth Sci Int 3(3):1–14

Saidou A (2006) Converging strategies by farmers and scientists to improve soil fertility and enhance crop production in Benin. Wageningen University and Research

Salick J, Byg A (2007) Indigenous peoples and climate change. Tyndall Centre

Salick J, Ross N (2018) Traditional peoples and climate change. In: Planning for climate change. Routledge, pp 160–166

Sharma A, Choudhury M (2021) Changing climate and depleting water resources in the mountains with a case study from the Himalayas. In: Water conservation in the era of global climate change. Elsevier, pp 411–436. https://doi.org/10.1016/B978-0-12-820200-5.00013-0

World resources 2005—the wealth of the poor (9 Jan 2005). World Resources Institute. https://www.wri.org/research/world-resources-2005-wealth-poor

Chapter 10
Sustaining Traditions, Nurturing Resilience: Community and Indigenous Approaches to Climate Challenges

Parnali Bhattacharya

Abstract In tracing the grappling contributors of the twenty-first century's climatic change and subsequent environmental devastation, the direction and pace of climatic challenges, it can be pointed out that the major contributors to the climatic challenge are anthropogenic activities of modernization and rapid urbanization. However, in the face of modernization, the sustainable and holistic lifestyle practices of lifestyles of several indigenous communities on the environment consist of conventional beliefs, cultural practices, a traditional tapestry of social structures, diversified livelihoods, and sacred sustainable practices that impart conservation and preservation of resources to a great extent. This said ecological knowledge repository of the indigenous communities, which is referred to as the traditional environmental knowledge (TEK), is quintessential in applying their techniques to reverse the impacts of global change directly benefitting the global climate conditions, can be focussed on as a methodology framework by studying through the climatic justice model.

Keywords Indigenous communities · Traditional ecological knowledge · Climate justice model · Sustainable practice

Introduction

Environmental hazards exert a direct effect on life systems through climatic devastations and patterns of climatic changes. Naturally, only the gradual extinction of life can be apprehended over time if environmental checks are not put in place. As per (the Environmental Protection Agency 2023), anthropic factors occupy a significant chunk of the factors responsible for aggravating climatic susceptibility. On that note, mitigation efforts to resist abrupt climatic changes are unavoidable and critical for survival. However, while anthropic factors contribute to the maximum amount

P. Bhattacharya (✉)
Department of Genetics and Plant-Breeding (Agriculture), University of Calcutta, Kolkata, West Bengal, India
e-mail: parnalee@gmail.com

M. Choudhury et al. (eds.), *Sustainable Synergy: Harnessing Ecosystems for Climate Resilience*, Climate Change Management,
https://doi.org/10.1007/978-3-031-77957-2_10

of climatic vulnerability, it is also the sustained anthropogenic efforts once again that can reverse the onset of climatic variabilities such as the traditional ecological concepts practised by several Indigenous communities of the world as a fitting answer to the pacing modernization (Yang et al. 2024).

The Climatic Justice Model serves as a great evaluator to study the TEKs of the global Indigenous communities through a detailed study of historical injustices if any, ecological vulnerability pointers, and promulgation of Indigenous rights and action on climatic benefit (Lefstad and Paavola 2023). Undoubtedly, the indigenous ethnic communities have been the most excellent umpires and arbiters of environmental conservation and sustainability. In this perspective, the communities' traditional ecological knowledge (TEK) emphasizes the holistic connection between humans and the environment, where the adjudication of climatic resilience forms a significant chunk. Thus, observing and subsequent rehabilitation of the indigenous communities and their holistic practices on culture, food procuring, and interdependence on ecosystems can act as an eye opener for the vulnerable areas of climatic changes and impart resilience by mitigating the effects of rapid climate change through climate justice pointers (Hill et al. 2024). Figure 10.1 shows pillars of climate justice.

Fig. 10.1 Pillars of climate justice

A Perspective on Indigenous and Ethnic Communities

It is the indigenous and the ethnic communities that are the first to incur the direct consequences of climate change and quite inherently mitigate the same with preparedness and re-building strategies due to their direct dependency and inter-relationship with natural resources and nature itself. Thus, the Indigenous and Local Knowledge (ILK) that can be sourced from various parts of the world from several ethnic communities must be studied through environmental assessment and practice models (Burgos-Ayala et al. 2020).

In this context, the Arctic Climate Impact Assessment (ACIA) and (Alexander et al. 2011) were the first massive systematic reports on climate migration resilience imparting and mitigation by indigenous communities. Since then, several reports of studying the traditional ecological knowledge (TEK) on extreme weather conditions and environmental hazard mitigation have been studied in several parts of the world with the help of several models fitting into the combined climatic justice. Some of the most common methods of the following include.

Community-Based Adaptation and Local Governance Model

In this approach, the needs, requirements, and priorities of the ethnic communities are mapped by observation and documentation in separate records. Based on these concomitants, an environment response action of a particular indigenous community is comprehended by the bottom-up participatory non-governmental organization. Then, a capacity assessment of the local governmental support network and other collaborations will be applied to mitigate a particular climatic problem. An example can be cited of the Peruvian Andes Quechua communities that have mitigated the impact of droughts and uneven rainfall through their traditional agricultural diversification technique (Feola 2019).

Indigenous Rights and Stewardship Model

As the name suggests, in this model, the respective Indigenous communities fight for their rights and justice system to protect their ecological resources, thereby securing an ecological balance against possible impending climatic devastation due to modernization. An example can be cited of the Kayapo community of Amazon, which has successfully resisted land rights in the face of deforestation (Rorato et al. 2021).

Bio-cultural Diversity and Resilience Model

In this concept, Indigenous communities are encouraged to observe biological, cultural, and socio-environmental holistic biodiversity-based practices to protect themselves from environmental alarms and hazards. The Zuni tribe's elaborate water management system can be cited as an example of this in the face of water scarcity due to climatic migration (Chief, Meadow, and Whyte 2016).

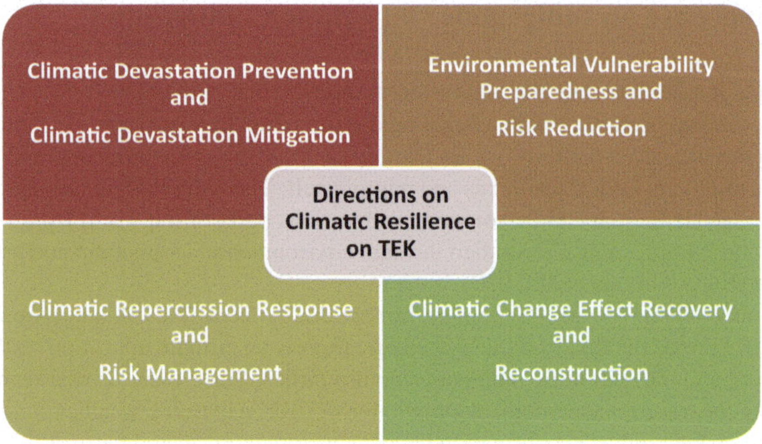

Fig. 10.2 Areas of application of traditional ecological knowledge (TEK)

Ecosystem-Based Adaptation Model

In this case, ecosystem conservation is encouraged to protect the topography and the existential environmental reaction of the particular area, restoring climatic impacts such as water availability, food security, and habitat restoration at the expense of modernization's intervention (Choudhury et al. 2021). The Indigenous communities of the Pacific Islands can be cited as an example (Mcleod et al. 2019).

Knowledge Gaps

Models, top-to-bottom participations, and inclusions of the traditional ecological knowledge of the Indigenous communities on environmental sustainability and climatic migration resilience continue to propagate. They can be easily studied through the TEK framework, Ethno ecological Framework, the (IPCCA), The Indigenous Peoples' Biocultural Climate Change Assessment Initiative, and many more. However, as per (Petzold et al. 2020), the lacuna that remains in exploring the extent of benefit is the lack of TEK and ILK practices or their sufficient patronizing in the scope of modernization. Thus, a systematic adaptation approach can be attempted from the exploratory research with the help of climate vulnerability, feasibility, and resilience index by focussing on relevant TEK practices by the ILP (Indigenous Local People). Figure 10.2 shows application of traditional ecological knowledge.

Methodology

From the study by (Buhaug et al. 2022), it has been found that the present century is grappling with magnanimous improvement in the quality of human life due to increased technological intervention; red signals of environmental disasters also flag

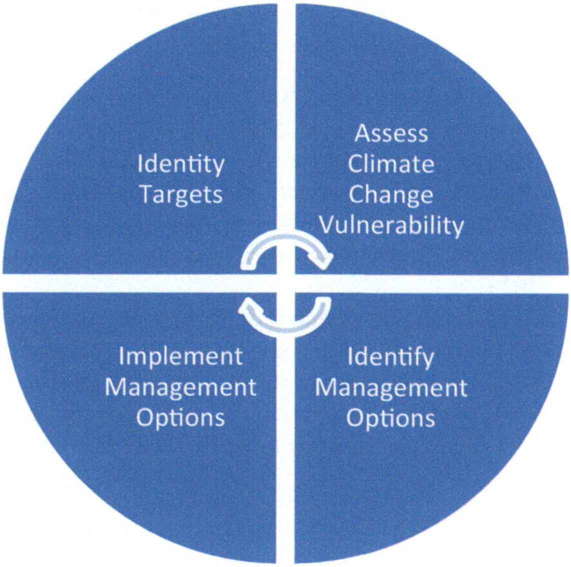

Fig. 10.3 Steps to implement TEK on climatic resilience

it, the trigger points of most of which are anthropic, and anthropogenic activities of development involving resource usage all contributing to global warming. Whether it is a flood, drought, environmental shift, demographic shift, food scarcity or gradual extermination of life forms, global warming is the foremost, massive climatic change whose offshoots are the given dangers (NASA 2024). Figure 10.3 shows steps to implement TEK on climatic resilience.

Traditional Knowledge Systems

The most significant contributor to global warming is the modernization of fossil fuels for modernization in agriculture and civilization (Jones et al. 2023). Thus, the TEKs must be directed and utilized to understand the suitable practice of civilization with the minimal usage of fossil fuels without deterring the properties of nature, which can only be achieved by a prior understanding of natural patterns of the environment and taking precautionary measures. Naturally, in this study, the objective will be to utilize all those aspects of TEK, where the practices are related to the following systems. Indigenous Knowledge (IK) is a set of cultural practices, ecological strategies, and adaptation methods to climatic migration. This knowledge is rooted in the close interdependence and connection of the indigenous communities with nature. Thus, the indigenous practices on climatic resilience are made up of experiences and observations that form the framework of adaptive strategies by the indigenous people (IP). On that note, the TEK, in this respect, can be defined as the indigenous knowledge system that combats climatic changes, especially those that are targetable towards global warming. The following sets of IK, with their impacts on climatic adaptation, can shape the climatic justice model.

i. **Weather Prediction IK**—Natural observations of natural resources, plant, and animal behaviour and planetary phenomena acts as a foremost indicator in weather predictions. Examples must be cited of; Zulu Traditional Weather Forecasting knowledge system, where the community predicts climatic trends through insect behaviour on flowering, and wind direction to understand the pattern of humidity, and manage their agriculture, forestry, and livelihood (Ubisi et al. 2023).
ii. **Agro-ecological behaviour of IP**—In this scenario, the ethnic communities, tend to understand the precipitation patterns and cloud cover through the movement of constellations (Hamacher et al. 2019). Worthwhile mentions must be made of Andean Community's agro-ecological practices based on constellation study to predict rainfall. Table 10.1 shows certain global indigenous community's contributions to climate resilience

Community-Based Approaches

The role of the indigenous communities in sustaining traditions and practices is mainly exemplified by the conservation of biodiversity, climatic relocation, land management, and resource usage. On that note, certain rituals practised by several Indigenous communities must be taken into consideration. In fact, to tally with the previous sentence's proposition, worthwhile mentions must also be made of specific community-led adaptation and mitigation projects as follows.

1. **Examples of community-led climate adaptation and mitigation projects**
   (a) Amazonian rainforest restoration by the Yawanawa tribe by planting native flora, sustainable land-use, and increasing carbon sequestration (Da Cruz et al. 2020).
   (b) Restoration of mangrove forests by the native community of the Gangetic delta, which has been an excellent buffer against cyclones, drought, and storms through absorption of blue carbon.
   (c) Sustainable soil management-based agricultural practices by the Mossi farmers of Sahel region of Africa, where the inclusion of stone buds, and zai pits, across farms have resulted in the retention of soil moisture (Kebenei et al. 2023).
   (d) The cool burns forest management technique of the aboriginal communities of Australia, where low intensity fire-stick farming and fuel management system has greatly contributed to reduction in greenhouse emissions (Williamson 2022).
2. **Benefits of community involvement and collective action**

Table 10.1 Certain global indigenous community's contributions to climate resilience

Geographical division	Tribe/community	Historical injustice on climate	Ecological vulnerability pointers	TEK (traditional ecological knowledge)	Climatic resilience output
Asia	Kalash (Chitral Valley, Pakistan)	Likely displacement and marginalization due to climate impacts and lack of recognition of their environmental role	Climate sensitivity due to reliance on agriculture and natural resources	Suri Jagek (traditional weather forecasting), community hydro-power, sustainable agriculture, forest and water conservation	Reducing emissions and reversing global warming through resource conservation and sustainable agriculture (World Bank Blogs, 15AD)
	Tsum (Nepal)	Deforestation and climate impacts threatening their biodiversity	Vulnerable to deforestation and biodiversity loss	Cultural patrolling, biodiversity preservation, and forest conservation	Preservation of biodiversity and reduction of emissions through cultural and ecological practices (World Bank Blogs, 15AD)
	Kondh (Orissa, India)	Threats from industrial activities (mining) impacting their environment and traditional lands	Vulnerability to loss of forest cover and biodiversity due to industrialization	Cultural patrolling, forest conservation, sustainable agriculture	Biodiversity expansion and climate resilience through forest conservation (World Bank Blogs, 15AD)
Africa	Sub-Saharan African Tribes	Climate change-induced desertification, food insecurity, and displacement	Vulnerability to drought and loss of ecosystems	Flexibility in agriculture, maintaining diverse food ecosystems	Reducing long-term risk to ecosystems and people, climate resilience through diversified food systems (Adeola et al. 2023)
	Ubuntu Tribes	Historical exploitation of natural resources and exclusion from conservation policies	Vulnerability to ecosystem degradation and biodiversity loss	Ubuntu networking for passing down traditional knowledge, promoting biodiversity preservation	Resilience through biodiversity prevention and long-term ecosystem protection (Adeola et al. 2023)

(continued)

Table 10.1 (continued)

Geographical division	Tribe/community	Historical injustice on climate	Ecological vulnerability pointers	TEK (traditional ecological knowledge)	Climatic resilience output
East Pacific	Maoris (Pacific Islands)	Sea-level rise and habitat loss due to greenhouse gas emissions	Vulnerability to rising sea levels, loss of biodiversity	Aforestation with local biodiversity, ecosystem preservation	Resilience against climate change through forestation and biodiversity conservation (Park et al. 2023)
	Tongans (Pacific Islands)	Sea-level rise and intensifying storms threatening their homes and traditional ways of life	Vulnerability to climate-induced disasters and biodiversity loss	Aforestation with local biodiversity, ecosystem preservation	Climate resilience through local biodiversity and reduced greenhouse gas emissions (Park et al. 2023)
	Samoans (Pacific Islands)	Threatened by sea-level rise and increasing climate variability	Vulnerability to habitat loss due to climate-induced changes	Ecosystem preservation and aforestation with native species	Climate resilience and reduced greenhouse gases through traditional ecological practices (Park et al. 2023)
	Hawaiians (Pacific Islands)	Climate change impacts on land, oceans, and traditional food systems	Vulnerability to habitat destruction and loss of biodiversity	Aforestation, biodiversity conservation, and traditional ecosystem management	Reduced vulnerability through ecosystem preservation and local reforestation efforts (Park et al. 2023)
Latin America and Caribbean	Guarayo (Latin America)	Historical displacement and loss of land due to industrial agriculture and deforestation	Vulnerability to land degradation and climate-induced relocation	Climate-induced relocation, aforestation, sustainable resource management	Vulnerability prevention, climate resilience through relocation, and sustainable land use (Felipe Pérez and Tomaselli 2021)

(continued)

Table 10.1 (continued)

Geographical division	Tribe/community	Historical injustice on climate	Ecological vulnerability pointers	TEK (traditional ecological knowledge)	Climatic resilience output
	Quechua (Latin America)	Historical marginalization, climate-induced habitat loss	Vulnerability to habitat degradation, biodiversity loss, and extreme weather events	Climate-induced relocation, traditional knowledge of land management and resilience-building	Vulnerability mitigation and preparedness through sustainable land use and resource management (Felipe Pérez and Tomaselli 2021)
	Baure (Latin America)	Threatened by climate change impacts on biodiversity and natural resources	Vulnerability to extreme weather events, habitat loss, and biodiversity decline	Traditional methods of climatic risk management and climate resilience response	Climate effect recovery and desired reconstruction through the application of traditional knowledge in land management and sustainable practices (Felipe Pérez and Tomaselli 2021)

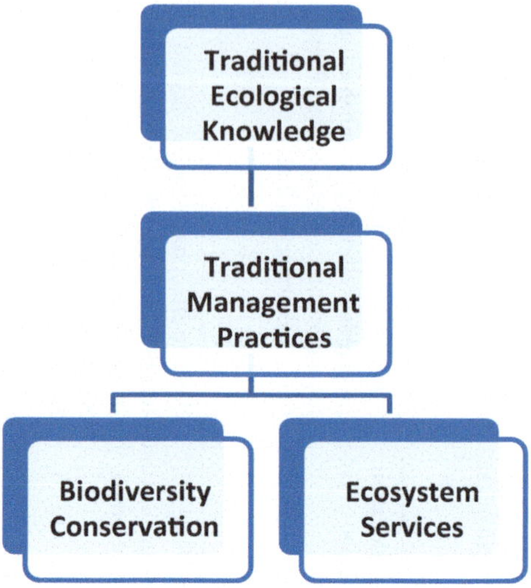

Fig. 10.4 Major avenues of indigenous community TEK contributions in climatic resilience

(a) Through traditions—In this respect, the controlled fishery tradition of the indigenous tribes of the Solomon Islands deserves a special mention that has greatly acted on preservation of coral reefs (Albert et al. 2015).
(b) Environmental Stewardship—The responsible uses and responsibilities on resource utilization has been beautifully exemplified by Ojibwe people of North-America, who by reintroduction of diverse species of rice in the rice fields, thereby making them a carbon sink and imparting effect of reduction of environmental temperature (Pimbert et al. 2021).
(c) Role of rituals, ceremonies, and storytelling in preserving knowledge activities of the Ashanti tribe of Ghana must be given a special mention where their traditional knowledge on environmental protection is passed on to generations thereby encouraging the participation of the youths in the act of imparting climatic resilience (Asante et al. 2017). Figure 10.4 shows major avenues of indigenous community.

Results

As urbanization generates the most anthropogenic activities, the climatic justice model states the need for responsibility equity, human rights emphasis, fair transition, and prioritization of environmental resilience policies (Strange et al. 2024). Whenever policies are considered, it is imperative to note the respective milestones that make up a policy framework that can connect the contributions of Indigenous communities, for which the barriers to implementing Indigenous practices to climate change must be considered (Kumar and Choudhury 2021).

Challenges and Barriers

The contemporary challenges faced by Indigenous communities in the context of climate change can be summarized as external pressures mainly marked by political, economic, and climatic impacts as a repercussion of modernization, whereas the internal pressures have a direct impact on the Indigenous community's TEKs are marked by; generational population shifts, and loss of traditional knowledge. In this aspect, the mention must be made of the challenges faced by the Inuits of the Arctic region, resulting in the rapid loss of traditional knowledge on ice patterns, and climatic temperature patterns (Ready and Collings 2020).

Policy and Advocacy

Thus, policy support for Indigenous Knowledge and practices is tremendously important through recommendations for future policy development. Protecting Indigenous Knowledge on climate resilience largely depends on advocating Indigenous rights through science-policy negotiations involving biodiversity and cultural traditions (Robinson et al. 2021). The Treaty of Waitangi Act 1975 deserves special mention because it has greatly advocated the rights of the Maori tribes regarding their participation and mutual benefit from the government regarding the traditional ecological Knowledge of climatic resilience and adaptation practices.

Collaborative Efforts

Naturally, the role of partnerships in enhancing climate resilience as lessons learned from collaborative projects emphasizing the complete consultation and participation of the indigenous communities in mainstream environmental justice. Worthwhile mentions must be made of Canada's Indigenous Climate Leadership Initiative programme, mainly focussing on including the chunk of local tribal communities for the protection of water resources as a measure of climatic resilience (Champagne-Côté et al. 2023) (Fig. 10.5).

Discussion

Applying the scenario of the climatic justice model comes hand in hand with the possibility of adaptation planning in the face of urbanization and modernization. The amalgamation of the following steps ensures the complete inclusion of TEK

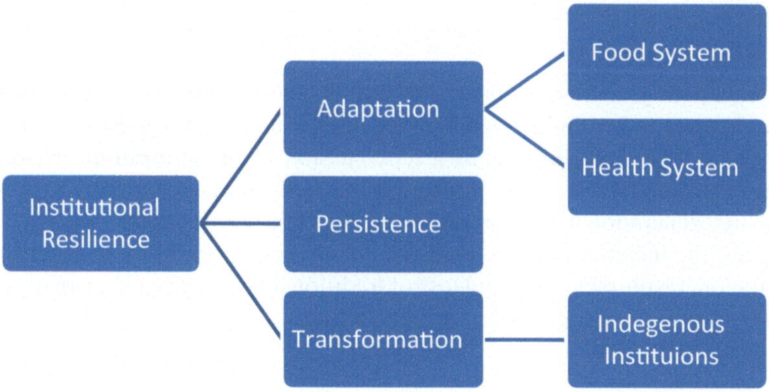

Fig. 10.5 Climatic resilience rights and policy advocacy example (Champagne-Côté et al., 2023)

from a selected Indigenous community of choice based in a particular region with the desired effect to pave the way for future directions (Juhola et al. 2022).

- **Calculation of Climatic Vulnerability Index**—This tool or scale helps to locate the hot spots and places of estimated climatic migration by mapping the level of exposure, and sensitivity.
- **Steps of choice detection**—In this stage, the selection criteria of the Indigenous community's practices are based on the calculation of;
  - **Threshold Capacity**—Threshold Capacity, or Carrying Capacity is the amount of upper and lower limit of sustained resources at particular conditions of temperature, and pressure. A pattern of several threshold capacities help to identify the expected amount of resources in several hot-spots of global warming.
  - **Coping Capacity, Transformative Capacity, or Adaptive Capacity**—Ability of the entities to manage TEK strategy in terms of money that can be systematically arranged to yield a desired result.
  - **Recovery Capacity**—The total number of resources such as biodiversity that can be reversed or reclaimed by a selected TEK strategy in terms of per unit benefit derived from per unit money invested for reversing a particular symptom of climate change.
- **Resilience Index, or Bioclimatic Ecosystem Resilience Index (BERI)**—Here the extent of a particular TEK on a particular geographical area is taken into consideration and the resilience index is expressed as the amount of climatic shift that can be mitigated by a particular TEK.
- **Feasibility Index**—It is the compliance measurement of an adaptive strategy received from a TEK that determines the amount of climatic resilience in-terms of desired climatic factors on the face of urbanization.

Conclusion

The face of traditionalism in devising a strategy for developing climate change resilience through sustainable principles may vary as per the diversity of the mapped requirements of a particular geographical region. However, utilizing the Indigenous communities requires a vast devotion to watching over their response to climatic threats before trying to grasp their traditional ecological knowledge straightaway. As per the study conducted, it is evident that the likeable direction of adopting an Indigenous community's climate resilience practices requires careful observation of the statistical results of calculating and tallying the vulnerability index of a practice, feasibility index of adaptation, and final resilience index in years. A suitable alignment with several studies and analyzing a practice can help modernization steer away from climatic migration and develop resistance to its massively by developing the conditions of proportionality between requirement goal setting and holistic biodiversity maintenance.

References

Adeola O, Evans O, Ngare I (2023) African indigenous knowledge and climate change mitigation: towards an Afro-sensed perspective. In: Sustainable development goals series, pp 169–191. https://doi.org/10.1007/978-3-031-40124-4_8

Albert JA, Olds AD, Albert S, Cruz-Trinidad A, Schwarz A-M (2015) Reaping the reef: provisioning services from coral reefs in Solomon Islands. Mar Policy 62:244–251. https://doi.org/10.1016/j.marpol.2015.09.023

Alexander C, Bynum N, Johnson E, King U, Mustonen T, Neofotis P, Oettlé N, Rosenzweig C, Sakakibara C, Shadrin V, Vicarelli M, Waterhouse J, Weeks B (2011) Linking indigenous and scientific knowledge of climate change. BioScience [online] 61(6):477–484. https://doi.org/10.1525/bio.2011.61.6.10

Asante EA, Ababio S, Boadu KB (2017) The Use of indigenous cultural practices by the Ashantis for the conservation of forests in Ghana. SAGE Open 7(1):215824401668761. https://doi.org/10.1177/2158244016687611

Buhaug H, Benjaminsen TA, Gilmore EA, Hendrix CS (2022) Climate-driven risks to peace over the 21st century. Clim Risk Manage 39:100471. https://doi.org/10.1016/j.crm.2022.100471

Burgos-Ayala A, Jiménez-Aceituno A, Torres-Torres AM, Rozas-Vásquez D, Lam DPM (2020) Indigenous and local knowledge in environmental management for human-nature connectedness: a leverage points perspective. Ecosyst People 16(1):290–303. https://doi.org/10.1080/26395916.2020.1817152

Champagne-Côté R, Beaudoin J-M, Bélanger L, St-Onge MR, Asselin H, Suffice P (2023) Indigenous leadership in creating a protected area: the Akumunan Biodiversity Reserve (Canada). Glob Ecol Conserv, e02681–e02681. https://doi.org/10.1016/j.gecco.2023.e02681

Chief K, Meadow A, Whyte K (2016) Engaging southwestern tribes in sustainable water resources topics and management. Water 8(8):350. https://doi.org/10.3390/w8080350

Choudhury M, Sharma A, Singh P, Kumar D (2021) Impact of climate change on wetlands, concerning Son Beel, the largest wetland of North East, India. In: Global climate change. Elsevier, pp 393–414. https://doi.org/10.1016/B978-0-12-822928-6.00006-X

Da Cruz DC, Benayas JMR, Ferreira GC, Santos SR, Schwartz G (2020) An overview of forest loss and restoration in the Brazilian Amazon. New Forests 52(52). https://doi.org/10.1007/s11056-020-09777-3

Environmental Protection Agency (2023) Causes of climate change [online]. www.epa.gov. Available at: https://www.epa.gov/climatechange-science/causes-climate-change

Felipe Pérez B, Tomaselli A (2021) Indigenous Peoples and climate-induced relocation in Latin America and the Caribbean: managed retreat as a tool or a threat? J Environ Stud Sci 11(3):352–364. https://doi.org/10.1007/s13412-021-00693-2

Feola G (2019) Climate and culture: multidisciplinary perspectives on a warming world. Cambridge University Press, ISBN-10. 1108422500; ISBN-13. 978-1108422505

Hamacher DW, Barsa J, Passi S, Tapim A (2019) Indigenous use of stellar scintillation to predict weather and seasonal change. Proc R Soc Victoria [online] 131(1):24. https://doi.org/10.1071/rs19003

Hill, K.X., Lyla June Johnston, Blue, M.R., Probst, J., Staecker, M. and Jennings, L.L. (2024). Rematriation and climate justice: Intersections of indigenous health and place. *The Journal of Climate Change and Health, 18*(100314). https://doi.org/10.1016/j.joclim.2024.100314

Jones MW, Peters GP, Gasser T, Andrew RM, Schwingshackl C, Gütschow J, Houghton RA, Friedlingstein P, Pongratz J, Le Quéré C (2023) National contributions to climate change due to historical emissions of carbon dioxide, methane, and nitrous oxide since 1850. Scientific Data [online] 10(1). https://doi.org/10.1038/s41597-023-02041-1

Juhola S, Heikkinen M, Pietilä T, Groundstroem F, Käyhkö J (2022) Connecting climate justice and adaptation planning: an adaptation justice index. Environ Sci Policy 136:609–619. https://doi.org/10.1016/j.envsci.2022.07.024

Kebenei MC, Mucheru-Muna M, Muriu-Ng'ang'a F (2023) The potential of Zai pit technology and Integrated soil fertility management to enhance crop productivity in semi-arid regions of Sub-Sahara Africa: a review. Int J Environ Agric Biotechnol 8(1):063–073. https://doi.org/10.22161/ijeab.81.8

Kumar D, Choudhury M (2021) Recognizing economic values of wetland ecosystem services: a study of emerging role of monetary evaluation of Chandubi ecosystem and biodiversity. In: Wetlands conservation: current challenges and future strategies, pp 87–110. https://doi.org/10.1002/9781119692621.ch5

Lefstad L, Paavola J (2023) The evolution of climate justice claims in global climate change negotiations under the UNFCCC. Critical Policy Stud, 1–26. https://doi.org/10.1080/19460171.2023.2235405

Mcleod E, Bruton-Adams M, Förster J, Franco C, Gaines G, Gorong B, James R, Posing-Kulwaum G, Tara M, Terk E (2019) Lessons from the Pacific Islands—adapting to climate change by supporting social and ecological resilience. Front Mar Sci 6(289). https://doi.org/10.3389/fmars.2019.00289

NASA (2024) The causes of climate change. [online] science.nasa.gov. Available at: https://science.nasa.gov/climate-change/causes/

Park S, Owens K, Summerfield-Ryan O, Taylor M, Ulloa AM, Viney G (2023) How can Pacific island countries meet their nationally determined contributions? NPJ Climate Action 2(1). https://doi.org/10.1038/s44168-023-00059-0

Petzold, J., Andrews, N., Ford, J.D., Hedemann, C., & Postigo, J.C. (2020). Indigenous knowledge on climate change adaptation: A global evidence map of academic literature. *Environmental Research Letters, [online] 15*(11), 113007. https://doi.org/10.1088/1748-9326/abb330

Pimbert MP, Moeller NI, Singh J, Anderson CR (2021) Agroecology. Oxford Res Encycl Anthropol. https://doi.org/10.1093/acrefore/9780190854584.013.298

Ready E, Collings P (2020) All the problems in the community are multifaceted and related to each other': Inuit concerns in an era of climate change. Am J Hum Biol 33(4). https://doi.org/10.1002/ajhb.23516

Robinson JM, Gellie N, MacCarthy D, Mills JG, O'Donnell K, Redvers N (2021) Traditional ecological knowledge in restoration ecology: a call to listen deeply, to engage with, and respect indigenous voices. Restoration Ecol 29(4). https://doi.org/10.1111/rec.13381

Rorato AC, Picoli MCA, Verstegen JA, Camara G, Silva Bezerra FG, Escada MIS (2021) Environmental threats over Amazonian indigenous lands. Land 10(3):267. https://doi.org/10.3390/land10030267

Strange KF, Satorras M, March H (2024) Intersectional climate action: the role of community-based organisations in urban climate justice. Local Environ, 1–21. https://doi.org/10.1080/13549839.2024.2315992

Ubisi N, Kolanisi U, Jiri O, Tshitangano F (2023) Spatial distribution of indigenous climate indicator development for rural smallholder farmers in Nkomazi local municipality, South Africa. Clim Change 176(9). https://doi.org/10.1007/s10584-023-03536-x

Williamson B (2022) Cultural burning and public forests: convergences and divergences between Aboriginal groups and forest management in south-eastern Australia. Aust for 85(1):1–5. https://doi.org/10.1080/00049158.2022.2054134

Yang H, He J, Li Z, Su Y, Xu J (2024) Ethnic diversity and divergent perceptions of climate change: a case study in Southwest China. Hum Soc Sci Commun 11(1). https://doi.org/10.1057/s41599-024-03207-x

Chapter 11
Resilience in the Digital Age: Emerging Technologies for Climate Adaptation

Laxmi Kant Bhardwaj, Pankaj Kanti Jodder, Rupam Priya, Prangya Rath, Harshita Jain, Shalini Thakur, Poonam Yadav, Sanju Purohit, and Bhawana Sharma

Abstract This chapter examines how breakthrough technology can address climate change and promote sustainability. It begins by reviewing climate change and emphasizing the need for prevention. According to the chapter, improved materials, blockchain, and artificial intelligence may be used in climate adaptation. These technologies can boost agricultural yields, reduce water consumption, and boost food security in a changing climate. Precision agriculture, hydroponics, and vertical farming may transform agriculture and ensure sustainable food production. Energy companies emphasize intelligent grids, energy storage, and renewable energy technology as key to a low-carbon economy and energy resiliency. These technologies help reduce greenhouse gas emissions and provide a sustainable energy future by switching from fossil fuels to renewable energy. It also monitors and manages climate-related threats using digital platforms, big data analytics, and IoT

L. K. Bhardwaj (✉)
Amity Institute of Environmental Toxicology, Safety & Management, Amity University, Noida, Uttar Pradesh, India
e-mail: bhardwaj.laxmikant@gmail.com

P. K. Jodder
Department of Geography and Earth Science, University of North Carolina at Charlotte, Charlotte, USA

R. Priya · B. Sharma
Department of Bioscience and Biotechnology, Banasthali Vidyapith, Vanasthali, Rajasthan, India

P. Rath · H. Jain
Amity Institute of Environmental Sciences, Amity University, Noida, Uttar Pradesh, India

S. Thakur
Circular Materials, St. Clair Ave. West, Toronto, ON M4V 1K6, Canada

P. Yadav
ICAR-Indian Institute of Maize Research, Ludhiana, Punjab, India

S. Purohit
Environmental/Ecological Studies and Sustainability, Akamai University, Kamuela, HI 96743, USA

M. Choudhury et al. (eds.), *Sustainable Synergy: Harnessing Ecosystems for Climate Resilience*, Climate Change Management,
https://doi.org/10.1007/978-3-031-77957-2_11

devices. These tools simplify adaptive decision-making, early warning, and community resilience. According to the chapter, enterprises, governments, and communities must work together to use new technology efficiently. It emphasizes the importance of inclusiveness, data privacy, and ethics when employing new climate adaptation technology. The chapter finishes by highlighting the transformative potential of new technologies to create a sustainable future and calling for coordinated efforts to maximize their benefits and resolve related concerns.

Keywords Climate change · Sustainability · Climate adaptation · Water consumption

Introduction

Urgent and creative steps are required to limit the impacts of the changing climate and guarantee a long-term sustainable future for subsequent generations. Effective adaptive solutions are becoming increasingly necessary as human activities negatively affect the Earth's climate. With their unparalleled potential to tackle the complex issues brought about by climate change, emerging technologies provide a glimmer of hope. This chapter explores how innovations across several industries might fundamentally alter our perspective on environmental sustainability and resilience, delving into the transformative potential of cutting-edge technologies in climate adaptation (Egerer et al. 2021).

Climate alteration is predominantly driven by greenhouse gas emissions, largely resulting from human activities such as the widespread use of fossil fuels and deforestation. As a result, it has exacerbated several negative repercussions, like increasing sea levels, severe weather phenomena, and rising global temperatures (Bhardwaj 2023; Bhardwaj et al. 2023). These effects put food security, human health, and socioeconomic consistency in danger, endangering ecosystems and biodiversity (Kumar et al. 2020). With the situation's urgency, emerging technologies become essential weapons in the battle against climate change. Advanced materials, blockchain, artificial intelligence, and renewable energy sources are just a few of these technologies' possible uses in climate adaptation. Smart grids, IoT-enabled monitoring systems, precision agriculture, vertical farming, and other technical innovations present revolutionary opportunities for building a more robust and sustainable future to climate change (Airoldi et al. 2021).

Innovative techniques like precision agriculture, hydroponics, and vertical farming provide viable answers to maximize output while reducing environmental effects when changing climatic patterns threaten conventional farming methods. Farmers can maximize resource allocation and reduce risks related to climate unpredictability by initiating automation technology. Additionally, the application of blockchain technology may enhance the agricultural supply chain's traceability and openness, promoting increased resilience and sustainability through the food production process (Herman et al. 2020). Innovations in producing sustainable energy, including solar and wind power, also provide scalable answers to the world's expanding energy needs while reducing carbon emissions and environmental damage (Lin et al. 2021).

Digital technologies are revolutionizing industries like agriculture and energy and are essential for improving adaptable and resilient decision-making to climate-related risks. We can monitor and manage climate-related hazards in real time using digital platforms, big data analytics, and IoT devices. Digitization promotes community empowerment and involvement, allowing for the collaborative development of robust infrastructure and adaptive tactics specific to the local community's requirements and vulnerabilities (Kang et al. 2020). The successful implementation of cutting-edge technologies for climate adaptation requires cooperation and inclusion among all sectors and stakeholders. Coordinated efforts are needed to fully utilize technology advancements while addressing ethical issues and protecting data privacy, from enterprises and governments to communities and civil society organizations. Furthermore, emerging technologies should be developed and implemented under the guidance of ethical standards and legal frameworks to ensure that they maximize benefits to society and minimize hazards and unexpected outcomes (Khan et al. 2020).

Climate Change: Issues and Need for Immediate Action

In the intricate tapestry of modern global challenges, climate change emerges as a paramount concern, casting a long shadow over the sustainability of the ecosystems of our world and the future of human civilization. This multifaceted phenomenon, driven by the accretion of greenhouse gases in Earth's atmosphere, is a testament to the profound impact of human activity on the natural world. This warming is not an abstract statistical anomaly but a harbinger of a cascade of environmental perturbations, each more complex and unpredictable than the last. These changes herald a future when extreme weather events increase severely, sea levels engulf coastal communities, and biodiversity suffers unprecedented losses (IPCC 2022).

The implications of climate change extend far beyond the environmental sphere, insinuating themselves into every facet of human life. The agricultural sector faces the specter of declining yields as crops falter under the stress of altered rainfall patterns and extreme temperatures (Haldar et al.2023). Water scarcity, exacerbated by diminishing glaciers and erratic precipitation, threatens to become a perennial challenge, compelling communities to contend with the daunting prospect of securing a vital resource. This crisis is underpinned by a complex web of causes, which include combustion in industrial operations, deforestation, and the use of fossil fuels. The challenge of climatic change is thus not merely environmental but intrinsically linked to the very essence of human progress and survival (Cooper and White 2022).

Urgency permeates the imperative to address climate change as temperatures soar and extreme weather events escalate, necessitating swift, unified global action. Transitioning to renewable energy sources is paramount, supported by policies aligned with scientific findings and emphasizing sustainability. Enhancing resilience entails robust infrastructure and disaster preparedness measures, particularly in vulnerable developing countries. Education campaigns are crucial in fostering public support for climate action, while policy reforms such as carbon pricing and green subsidies incentivize sustainable choices. A comprehensive, collaborative effort is imperative to mitigate these effects to ensure everyone has a sustainable future (Chou et al. 2023).

Emerging Technologies: A Revolutionary Method

Emerging technologies are increasingly becoming a cornerstone in the global strategy for climate adaptation, offering novel solutions that revolutionize how societies prepare for and address the issues brought about by climate change. This transformation is rooted in enhancing and applying various technologies, including blockchain, sophisticated materials science, artificial intelligence (AI), and the Internet of Things (IoT).

Artificial Intelligence and Machine Learning stand at the forefront of climate adaptation technologies. Their capacity to process and analyze vast amounts of data from climate models, satellite imagery, and historical weather patterns enables precise predictions of future climate events (Galaz et al. 2021). This predictive power is invaluable for developing early warning systems that can lessen financial losses and save lives by preparing communities for extreme weather events like hurricanes, floods, and heat waves. Furthermore, AI is instrumental in optimizing renewable energy sources. By predicting wind patterns and sunlight exposure, AI algorithms can significantly enhance the efficiency of wind farms and solar panels, making renewable energy more viable and accessible.

The Internet of Things (IoT) offers a dynamic framework for climate adaptation through its network of interconnected devices. In agriculture, IoT sensors can transform traditional practices into smart farming techniques by monitoring temperature, moisture content of the soil, and crop health in real time (Rajak et al. 2023). This approach not only conserves water and increases crop resilience but also reduces the environmental footprint of agricultural practices. Moreover, deploying IoT sensors across diverse ecosystems provides critical data on environmental changes, facilitating more informed decisions for conservation efforts and climate adaptation strategies. IoT technologies enable real-time monitoring and data collection to enable societies to respond more effectively to the changing climate (Mondejar et al. 2021).

Blockchain technology introduces a layer of transparency and security that is particularly beneficial for climate adaptation efforts. Its application in tracking carbon emissions and trading carbon credits can revolutionize climate finance by ensuring that carbon offsetting efforts are accurately recorded and verifiably impactful. Additionally, blockchain can enhance the distribution of resources in disaster relief scenarios by creating transparent and efficient channels for aid, ensuring that help reaches those in need promptly and without loss through corruption or inefficiency (Kim and Huh 2020).

Developing advanced materials is crucial for constructing infrastructure capable of withstanding extreme weather conditions. Innovative materials such as self-healing concrete, corrosion-resistant alloys, and water-repellent surfaces pave the way for buildings and roads that last longer and require less maintenance, even in harsh climate conditions. Furthermore, advances in materials science are leading to the creation of more efficient insulation materials, reducing energy consumption in buildings by contributing to the worldwide initiative to lessen climate change (Dalawai et al. 2020).

Agriculture: Strengthening Sustainability and Resilience

The leading problems brought on by changing weather patterns, water scarcity, environmental degradation, and agriculture are the leading problems caused by changing climate (Singh et al. 2022). The agricultural industry, essential to rural livelihoods and global food security, needs to quickly adjust to lessen these effects and maintain sustainable production methods (Osaka et al. 2021). A paradigm change in agricultural operations, precision agriculture uses technology to customize crop management plans to fit the particular needs of every plant and land area. Agriculturalists can monitor the amount of moisture, nutrients, and insect infestations in real time by integrating data from sensors, satellite imaging, and IoT devices. This allows for tailored interventions to maximize resource efficiency and reduce environmental impact. Farmers can enhance efficiency, reduce waste, boost yields, and enhance resilience to climate-related issues by precisely applying inputs such as water, fertilizer, and insecticides. Moreover, precision farming enables farmers to make adaptive decisions, enabling them to anticipate environmental changes and reduce the hazards brought on by climatic unpredictability (Hafeez et al. 2023).

Hydroponic systems reduce water consumption, do not require chemical fertilizers, and lessen soil degradation linked to conventional farming methods by cultivating plants without soil in nutrient-rich water solutions. Similarly, vertical farming uses stacked shelves or growing trays to optimize space utilization, allowing for high-density production in places with limited arable land or urban areas. These methods protect crops from extreme weather and temperature swings, lessening agriculture's environmental impact and increasing their resilience to climate change (Semeraro et al. 2023).

Blockchain technology allows farmers, producers, and consumers to trace agricultural products from farm to fork, guaranteeing sustainability, safety, and authenticity. Transactions are documented in a secure, immutable digital ledger. This transparency builds trust among stakeholders and incentivizes producers to adhere to ethical and environmental standards, thereby promoting sustainable practices. These advancements, which range from blockchain-enabled traceability systems to precision agriculture and hydroponics, enable farmers to reduce risks, maximize resource use, and adjust to shifting environmental circumstances (Balasundram et al. 2023).

Energy: Moving Toward a Future with Less Carbon

By 2030, 65% of the world's electrical supply may come from inexpensive renewable energy sources. In 2050, it might contribute towards 90% of carbon-free power supply by the industries, radically mitigating carbon emissions and supporting the battle against climate change. Biological resources have a large potential to provide the expanding energy demand. Decarburization and low carbon are essential to the energy transition and a practical means of reducing increased CO_2 emissions (Gogoi et al. 2024). Low carbon and decarburization are top priorities in international policy for more environmentally friendly energy supply solutions with improved energy efficiency. As Prasad et al. (2021) point out, bioenergy is crucial to the "low-carbon energy mix." The report concludes by identifying lower energy use, alternative renewable energy sources, and creating renewable energy. Central European countries can make the change to an economy based on sustainable energy and achieve economic and strategic carbon neutrality. Implementing three key strategies, supply diversification, energy conservation, and rapid adoption of renewable energy to replace fossil fuels, can facilitate the EU energy sector's decarbonization goals and enhance energy security (Rabbi et al. 2022). Transitioning towards a low-carbon future necessitates affordable options, with governments playing a crucial role in managing emissions, setting regulatory frameworks, and promoting clean energy initiatives. Emphasizing energy efficiency is pivotal and achievable through various measures like heat pumps and improved building standards (Corwin and Pankratz 2021).

Utilizing Digital Solutions for Managing Risks

Concerning rapid changes in climatic conditions, cities worldwide are confronting myriad challenges, including heightened exposure to climate-related disasters and escalating waste issues, all of which strain adaptive capacities. The production of greenhouse gases, particularly plastic, contributes significantly to emissions, with estimates indicating annual costs reaching billions of dollars (Landrigan et al. 2023). Extreme weather events, increasingly influenced by climate change, have surged frequently, resulting in devastating floods in various regions globally. Digital technologies offer promising avenues to enhance risk assessments, support decision-making processes, and facilitate evidence-based approaches to integrate digitalization into future Sustainable Development Goals (SDGs) (Balogun et al. 2020;

Tokas et al. 2024). Leveraging digital solutions presents an opportunity to accelerate progress towards the Paris Agreement and SDG 13, with existing ICT solutions capable of potentially reducing current emissions by 15% (Bergmark 2018). Embracing digital transformation is imperative for businesses to thrive sustainably in the face of evolving trends like circular economy and servitization (El Hilali et al. 2020). "Industry 4.0" holds promise for industrial value creation but demands significant adaptations from companies, with potentially far-reaching impacts on sustainability (Tirabeni et al. 2019). As corporations navigate the virtual revolution, enhancing virtual capabilities and stabilizing economic, environmental, and social effects are paramount for long-term resilience (Birkel et al. 2019). Addressing the pressing consequences of changing climatic conditions requires ambitious emissions reduction targets, with commitments aiming for a 7.5% reduction by 2030. However, to align with the Paris Agreement's goals, a 55% reduction is imperative (Balaram 2023).

Working Together and Considering Ethics

Addressing climate change necessitates collaborative efforts across governments, businesses, NGOs, and individuals (IPCC 2022). Emerging technologies hold promise but require cooperative frameworks to be effectively leveraged, with public–private collaborations driving innovative solutions. Given the global nature of climate change, integration and cooperation across borders are imperative (IPCC 2022). Local communities play a crucial role in climate adaptation, requiring tailored solutions that engage community members in decision-making (Iizumi 2019). Ethical considerations must guide technology deployment, ensuring fair access, data security, and privacy protection (World Bank 2023). Technology should complement natural solutions, not replace them, emphasizing the importance of balancing technological advancements with nature conservation. Sustainability should drive technological innovation, prioritizing resource efficiency and greenhouse gas reduction (World Bank 2023). Education is vital in fostering awareness and accountability regarding technology's role in climate adaptation across all societal levels (IPCC 2022). Navigating the path to a resilient future requires careful consideration of sustainability, ethics, and teamwork, leveraging technology to create a sustainable and resilient future (IPCC 2022). As we collectively navigate this path, wise decisions, and collaboration will be essential in harnessing technology's potential for climate adaptation.

Opportunities and Challenges

The assimilation of digital technology offers unparalleled prospects for augmenting endeavors aimed at climate adaptation. Artificial intelligence, blockchain technology, and advanced materials provide creative ways to enhance resource efficiency, enhance resilience to climate-related disturbances, and enhance agricultural practices. Precision farming, hydroponics, and vertical farming have the potential to transform food production, maintain sustainability, and lessen the effects of climate change on the world's food supply. Similarly, decarbonizing the energy sector and improving energy resilience in the face of climate unpredictability is possible with the help of smart grids, energy storage systems, and renewable energy technology (El Hilali et al. 2020). Developing technology for climate adaptation is fraught with difficulties, notwithstanding their potential advantages. Uneven distribution of technology equipment and skills is a major barrier that can impede adoption and accessibility, especially in poor and marginalized populations. Furthermore, challenges of interoperability and scalability impede the broad adoption of technology solutions in various contexts. Moreover, it is still crucial to guarantee inclusivity and fair access to technology developments since gaps in digital literacy and access could make already-existing vulnerabilities and inequities worse (Landrigan et al. 2023).

Recommendation

Several crucial suggestions are made on how to use cutting-edge technologies best to combat climate change and advance sustainability. First and first, funds and resources for creative approaches to climate adaptation must be allocated to Research and Development (R&D) projects. Collaboration among academics, industry, and government bodies is imperative in propelling the advancement of these technologies. Secondly, governments should put supportive measures in place to promote the integration of cutting-edge technologies like blockchain, AI, and better materials into climate adaptation projects. This can entail providing rewards for experimental initiatives and efficacy demonstrations. Thirdly, cooperation between businesses, governments, and communities is crucial to fully realizing the potential of new technologies. To achieve shared objectives, partnerships should be established to pool varied resources and experience.

Furthermore, equity and inclusivity must be guaranteed when implementing these technologies, especially in marginalized groups. Strong ethical standards and data privacy laws are also essential for preventing the improper use of personal information. In addition, it is imperative to initiate education and awareness initiatives to enhance the general public's comprehension of the advantages and obstacles linked to these technologies. Achieving global sustainability targets also requires promoting international cooperation and scaling up successful initiatives.

Conclusion

Integrating emerging technologies into climate adaptation strategies are offers hope in the fight against climate change. By leveraging AI's predictive capabilities, IoT's connectivity, blockchain's transparency, and advancements in materials science, societies can bolster their resilience and adaptability to climate impacts. These technologies present innovative solutions to current challenges and open the door to a sustainable future. As these technologies evolve, their application in climate adaptation efforts will become more sophisticated, holding even greater potential for safeguarding the planet for future generations. This chapter provides insights into the intersection of advanced innovation and information systems in combating climate change, emphasizing the role of technology-based solutions and the importance of holistic educational initiatives to address e-waste concerns. Developed nations must prioritize efficient digitalization to achieve sustainable change, recognizing the obligation to devise solutions to their challenges. As innovation drives towards zero emissions by 2050, decision-makers must remain aware of these transformative options.

References

Airoldi L, Beck MW, Firth LB, Bugnot AB, Steinberg PD, Dafforn KA (2021) Emerging solutions to return nature to the urban ocean. Ann Rev Mar Sci 13:445–477

Balasundram SK, Shamshiri RR, Sridhara S, Rizan N (2023) The role of digital agriculture in mitigating climate change and ensuring food security: an overview. Sustainability 15(6):5325

Balaram V (2023) Combating climate change and global warming for a sustainable living in harmony with nature. J Geogr Res 6(3):1–17

Balogun AL, Marks D, Sharma R, Shekhar H, Balmes C, Maheng D, Arshad A, Salehi P (2020) Assessing the potentials of digitalization as a tool for climate change adaptation and sustainable development in urban centres. Sustain Cities Soc 53:101888

Bhardwaj LK (2023) A comprehensive review on the climate change and its impact on health. Preprints 2023:2023050159. https://doi.org/10.20944/preprints202305.0159.v1

Bhardwaj LK, Rath P, Yadav P (2023) A comprehensive review on the impact of climate-induced livelihood, health and migration on women and girls. Preprints 2023:2023122271. https://doi.org/10.20944/preprints202312.2271.v1

Birkel HS, Veile JW, Müller JM, Hartmann E, Voigt KI (2019) Development of a risk framework for Industry 4.0 in the context of sustainability for established manufacturers. Sustainability 11(2):384

Bergmark P (2018) Reflections regarding ICT and a citizen-centric future path of smart sustainable cities: Aw4city 2018 keynote. In: Companion proceedings of the web conference, pp 929–933

Chou C, Clark R, Kimbrough SO (2023) What do firms say in reporting on impacts of climate change? An approach to monitoring ESG actions and environmental policy. Corp Soc Responsib Environ Manag 30(5):2664–2678

Cooper N, White A (2022) IPCC report: urgent climate action needed to halve emissions by 2030. In: World economic forum, vol 6

Corwin S, Pankratz DM (2021) Leading in a low-carbon future: a "system of systems" approach to addressing climate change

Dalawai SP, Aly MAS, Latthe SS, Xing R, Sutar RS, Nagappan S, Ha C-S, Sadasivuni KK, Liu S (2020) Recent advances in durability of superhydrophobic self-cleaning technology: a critical review. Prog Org Coat 138:105381

Egerer M, Haase D, McPhearson T, Frantzeskaki N, Andersson E, Nagendra H, Ossola A (2021) Urban change as an untapped opportunity for climate adaptation. NPJ Urban Sustain 1(1):22

El Hilali W, El Manouar A, Idrissi MAJ (2020) Reaching sustainability during a digital transformation: a PLS approach. Int J Innov Sci 12(1):52–79

Galaz V, Centeno MA, Callahan PW, Causevic A, Patterson T, Brass I, Baum S, Farber D, Fischer J, Garcia D, McPhearson T, Jimenez D, King B, Larcey P, Levy K (2021) Artificial intelligence, systemic risks, and sustainability. Technol Soc 67:101741

Gogoi N, Choudhury M, Ul Hasan MS, Changmai B, Baruah D, Samanta P (2024) Soil organic carbon pool in diverse land utilization patterns in North-East India: an implication for carbon sequestration. Environ Dev Sustain, 1–27. https://doi.org/10.1007/s10668-024-05037-y

Hafeez U, Ali M, Hassan SM, Akram MA, Zafar A (2023) Advances in breeding and engineering climate-resilient crops: a comprehensive review. Int J Res Adv Agric Sci 2(2):85–99

Haldar S, Choudhury M, Choudhury S, Samanta P (2023) Trend analysis of long-term meteorological data of a growing metropolitan city in the era of global climate change. Total Environ Res Themes 7:100056. https://doi.org/10.1016/j.totert.2023.100056

Herman JD, Quinn JD, Steinschneider S, Giuliani M, Fletcher S (2020) Climate adaptation as a control problem: review and perspectives on dynamic water resources planning under uncertainty. Water Resour Res 56(2):e24389

Iizumi T (2019) Emerging adaptation to climate change in agriculture. In: Adaptation to climate change in agriculture: research and practices, pp 3–16. https://doi.org/10.1007/978-981-13-9235-1_1

Intergovernmental Panel on Climate Change (2022) Climate change: a threat to human wellbeing and health of the planet. Taking action now can secure our future. IPCC Reports

Kang JN, Wei YM, Liu LC, Han R, Yu BY, Wang JW (2020) Energy systems for climate change mitigation: a systematic review. Appl Energy 263:114602

Khan A, Gupta S, Gupta SK (2020) Multi-hazard disaster studies: monitoring, detection, recovery, and management, based on emerging technologies and optimal techniques. Int J Disaster Risk Reduct 47:101642

Kim SK, Huh JH (2020) Blockchain of carbon trading for UN sustainable development goals. Sustainability 12(10):4021

Kumar D, Choudhury M, Rathore A (2020) Valuation of ecosystem services & benefits of Son Beel wetland in Assam, India: a case study of natural solutions to climate change & water: the worldwide online climate change conference. Hamburg University, Germany

Landrigan PJ, Raps H, Cropper M, Bald C, Brunner M, Canonizado EM, Charles D, Chiles TC, Donohue MJ, Enck J, Fenichel P, Fleming LE, Ferrier-Pages C, Fordham R, Gozt A, Griffin C, Hahn ME, Haryanto B, Hixson R, Ianelli H, James BD, Kumar P, Laborde A, Law KL, Martin K, Mu J, Mulders Y, Mustapha A, Niu J, Pahl S, Park Y, Pedrotti ML, Pitt JA, Ruchirawat M, Seewoo BJ, Spring M, Stegeman JJ, Suk W, Symeonides C, Takada H, Thompson RC, Vicini A, Wang Z, Whitman E, Wirth D, Wolff M, Yousuf AK, Dunlop S (2023). The Minderoo-Monaco Commission on plastics and human health. Ann Glob Health 89(1)

Lin BB, Ossola A, Alberti M, Andersson E, Bai X, Dobbs C et al (2021) Integrating solutions to adapt cities for climate change. Lancet Planet Health 5(7):e479–e486

Mondejar ME, Avtar R, Diaz HLB, Dubey RK, Esteban J, Gómez-Morales A et al (2021) Digitalization to achieve sustainable development goals: steps towards a Smart Green Planet. Sci Total Environ 794:148539

Osaka S, Bellamy R, Castree N (2021) Framing "nature-based" solutions to climate change. Wiley Interdiscip Rev Clim Change 12(5):e729

Prasad S, Venkatramanan V, Singh A (2021) Renewable energy for a low-carbon future: policy perspectives. In: Sustainable bioeconomy: pathways to sustainable development goals, pp 267–284

Rabbi MF, Popp J, Máté D, Kovács S (2022) Energy security and energy transition to achieve carbon neutrality. Energies 15(21):8126

Rajak P, Ganguly A, Adhikary S, Bhattacharya S (2023) Internet of things and smart sensors in agriculture: scopes and challenges. J Agric Food Res 14:100776

Semeraro T, Scarano A, Leggieri A, Calisi A, De Caroli M (2023) Impact of climate change on agroecosystems and potential adaptation strategies. Land 12(6):1117

Singh S, Singh P, Sharma A, Choudhury M (eds) (2022) Agriculture waste management and bioresource: the circular economy perspective. Wiley. https://onlinelibrary.wiley.com/doi/book/10.1002/9781119808428

Tirabeni L, De Bernardi P, Forliano C, Franco M (2019) How can organisations and business models lead to a more sustainable society? A framework from a systematic review of the industry 4.0. Sustainability 11(22):6363

Tokas R, Bhardwaj LK, Kumar N, Jindal T (2024) Nanotechnology for sustainable development and future: a review. In: Green and sustainable approaches using wastes for the production of multifunctional nanomaterials, pp 221–233

World Bank (2023) Digital technologies fast track climate solutions. https://www.worldbank.org/en/news/press-release/2023/11/29/digital-technologies-fast-track-climate-solution

Chapter 12
Breaking Barriers: Innovative Approaches and Technologies for Climate Resilience: A Case Study from European Countries

N. Bharathy, P. Kowsalya, Einstein Mariya David, Moharana Choudhury, Parvez Alam Khan, Moses Kwame Aidoo, Saiyyeda Firdous, Eric D. van Hullebusch, and T. Parthasarathi

Abstract In recent decades, European countries have experienced significant consequences of climate change, including severe water scarcity and intense heat waves. In June 2024, the average temperature in these countries notably increased by 1.57 °C. This has further resulted in an unsustainable environment, leading to the extinction of natural resources, large wild fires, loss of biodiversity, decline in livestock population, and adverse effects on human well-being. To overcome these challenges, European countries are adopting innovative approaches and technologies, including nature-based solutions, smart cities and urban resilience strategies, carbon accounting technologies, ecosystem restoration efforts, intelligent water management systems, climate-resilient agricultural practices, and savvy digital solutions. In

N. Bharathy · P. Kowsalya · E. M. David · S. Firdous · T. Parthasarathi (✉)
VIT School of Agricultural Innovations and Advanced Learning, Vellore Institute of Technology (VIT), Vellore 632014, India
e-mail: parthasarathi.t@vit.ac.in

N. Bharathy · E. M. David
School of Bio Sciences and Technology, Vellore Institute of Technology (VIT), Vellore 632014, India

M. Choudhury
Environmental Research and Management Division, Voice of Environment (VoE), Guwahati, Assam, India

P. A. Khan
Department of Management and Humanities, Universiti Teknologi PETRONAS, Seri Iskandar, Malaysia

M. K. Aidoo
Cocoa Research Institute of Ghana, New Tafo-Akim, Eastern Region, Ghana

E. D. van Hullebusch
Université Paris Cité, Institut de Physique du Globe de Paris, CNRS, 75005 Paris, France

M. Choudhury et al. (eds.), *Sustainable Synergy: Harnessing Ecosystems for Climate Resilience*, Climate Change Management,
https://doi.org/10.1007/978-3-031-77957-2_12

addition to discussing these technologies, we will also explore the EU policy framework and public subsidies. This chapter covers a range of strategies and technologies implemented across European countries to reduce and adapt to the impacts of climate change. It also presents case studies featuring specific examples of European climate resilience projects. Furthermore, this chapter offers comprehensive insights into existing European technologies, enhancing global adaptability and laying the groundwork for developing innovative approaches to climate resilience worldwide.

Keywords Climate change · Resilient · Green infrastructure · Smart city · Restoration

Introduction

Recently, human-caused greenhouse gas emissions in the atmosphere, mainly from burning fossil fuels, industrial output, and agricultural operations, have contributed to global climate change (Kumar et al. 2021). Europeans are currently experiencing the consequences of global warming in their daily lives, which will continue into the years to come; severe water scarcity, heat waves, flooding, and other risks associated with climate change are becoming more apparent in Europe (European Environment Agency 2024). The rise in global climatic change influences everyone and is intensifying. Its effects will intensify if global temperatures exceed 1.5 °C (Hoegh-Guldberg and Bindi 2018). This has further resulted in an unsustainable environment, leading to natural resource depletion, ecological degradation, reduction in livestock population, and adverse effects on human well-being. To overcome these challenges, European countries are adopting novel approaches and technologies, including nature-based solutions, smart cities, urban resilience strategies, carbon counting technologies, ecosystem restoration efforts, intelligent water management systems, climate-resilient agricultural practices, and innovative digital solutions (Balogun et al. 2022; Calliari et al. 2022). In addition to discussing these technologies, we will also explore the EU policy framework and public subsidies. This chapter covers a range of strategies and technologies implemented across European countries to reduce and adapt to the impacts of climate change. It also presents case studies featuring specific examples of European climate resilience projects. Furthermore, this chapter offers comprehensive insights into existing European technologies, enhancing global adaptability and laying the groundwork for developing innovative approaches to climate resilience worldwide.

Technological Innovations for Climate Resilience

Nature-Based Technology and Green Infrastructure

Nature-based technologies (NbTs) are crucial in Europe's sustainable, regenerative, and resilient environments. NbT refers to practices inspired by Mother Nature, aimed at conserving environmental functionality or restoring ecosystems disrupted by anthropogenic activities (Kiss et al. 2022). This concept represents potential adaptation and resilience approaches with balanced confidence in socio-ecological domains (Dumitru and Wendling 2021). The European Union, recognizing the importance of NbTs, has recently launched HORIZON Europe, an innovation funding programme. This initiative is specifically designed to address disruptive climate changes, promote healthy soil and food systems, accelerate the transition to a sustainable well-being-oriented future, and achieve carbon-neutral, eco-smart cities by 2030 (European Commission 2018).

The 2021 Global Status Report reveals that the construction and operations of buildings were responsible for a significant 37% of global energy-related CO_2 emissions in 2020. To address this, implementing Nature-based Technologies (NbTs) is crucial. NbTs not only restore the water cycle, improve waste management, enhance food production, and optimize energy but also support building recovery (Calabrese et al. 2023). NbT in building systems can be categorized into three different tiers: green building materials, green building systems, and green building sites (Bona et al. 2022). The "City Park/Greening the City" project in Amsterdam is a shining example of successful NbT implementation. This project, which renovated green spaces, reduced flood risk, and strengthened political support and social awareness for water diplomacy and climate change adaptability, can serve as a model for similar initiatives (Carvalho et al. 2022; Mabon et al. 2022).

In cities like Berlin, Hamburg, and Dresden, Germany has implemented sustainable and socially integrative urban renewal projects, utilizing Nature-based Solutions (NbS) to develop green spaces. These projects, named BENE, Green Moabit, and Mischwald, have contributed to creating green corridors, roofs, facades, and school gardens. In Berlin, a greenbelt was established to enhance green connectivity, which increased urban resilience, biodiversity, and opportunities for social learning (proGIeg Project 2020). Similarly, cities of Italy, including Bari, Milan, and Turin, have adopted NbT initiative through projects and programs such as Operation Zero Degradation, Shagree: the Green Shadows program, Parco Agricolo Sud, Urban gardening Bosco Verticale, Gorla Maggiore Water Park, Rinverdiamo Milano, and the new park at Giambellino 129, as well as the Parco Sangone program (Arlati et al. 2021; Scharf et al. 2021). The primary objectives of these initiatives were to improve urban quality, reduce urban heat island effects, manage water resources, and create and strengthen green infrastructure, urban forestry, green roofs and walls, urban farming, and gardening projects (Fig. 12.1). These projects have significantly contributed to climate resilience by mitigating urban heatwaves, restoring and maintaining biodiversity and ecosystems, and regulating the water cycle. They promote the development of multipurpose green infrastructure, support recreational activities, reinforce public green spaces, and promote the innovative integration of green spaces into railway infrastructure. Furthermore, it supports entrepreneurship and green employment (Arlati et al. 2021; Oppla 2023; Scharf et al. 2021; Simperler et al. 2020). Figure 12.2 shows the schematic diagram of nature based technology.

Bristol City has adopted NbT technology in the United Kingdom to promote a sustainable future. The Green Infrastructure Green Areas project for flood management provided improved and protected green infrastructure, promoted healthy lifestyles, increased connectivity with nature, and fostered an inclusive social environment. As a result, the adverse effects of climate change were mitigated and

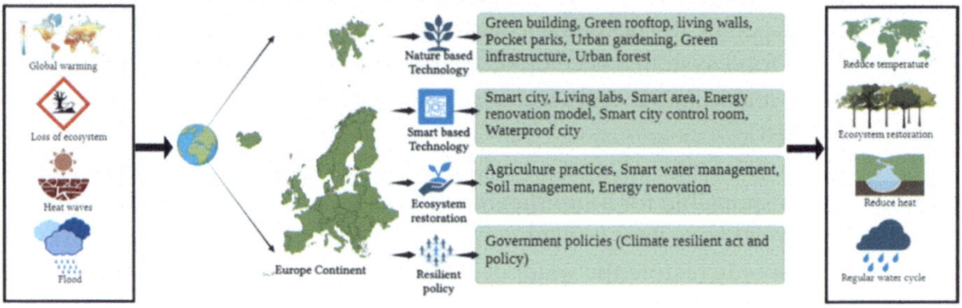

Fig. 12.1 Innovative approaches and technologies for climate resilience

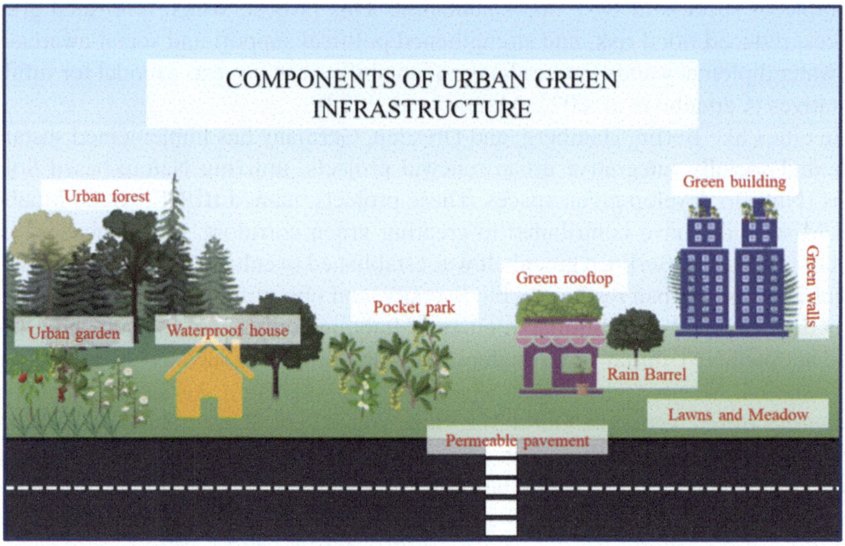

Fig. 12.2 Nature based technology

alleviated. Similarly, in Edinburgh, opportunities for new businesses and knowledge transfer, cultural wealth, green infrastructure connection, ecological diversity, cultural wealth, and climate adaption, public perception were achieved by the NbT approach through the Pollinator Pledge, Granton Community and Gardeners Duddingston Field Group project.

In London, NbT technology was applied in natural measures for retention at the River Quaggy, brownfield restoration at Barking Riverside, and the Olympic Park (Bio-solar Roofs and Beetle Bump) at the University of East London campus. These initiatives aimed to reduce heat stress, flood risk, and air quality issues (Clever Cities 2024; Mahmoud et al. 2021; O'Keeffe et al. 2022).

The programme "Green Area Rehabilitation and Green Infrastructure Development Urban Gardening" are in Szeged, Hungary, focused on restoring natural habitats and improving the quality of green spaces and environmental corridors for social and recreational reasons. This initiative creates more recreational environments and raises public awareness and ecosystem stability (Koppelaar et al. 2021; Marvuglia et al. 2020). Similarly, Poznan, Poland, has maintained a green wedge system by planting 18,000 trees along roadsides and incorporating transitional green elements. Furthermore, car parks have been converted into green spaces and riverside areas with seasonal beaches for leisure, creating high tourist potential and new home projects near green areas (Dymek et al. 2021; Zwierzchowska et al. 2021). In Romania, the city of Oradea has improved the quality of life of people by using NbT systems. They established the active use of newly created spaces with better recreation, connectivity, biodiversity, air quality, and climate conditions; the development of green infrastructure and green area rehabilitation; and new job and business initiatives, including lakes (Oppla 2023). An example of such initiatives is Dublin, the capital of Ireland, which launched the "NbT for a More Sustainable City by 2030" project. The project aimed to establish a sustainable and resilient city that considers the economy, environment, and equity. It provides an attractive place to live, work, and visit by enhancing air and water regulation, promoting climate adaptation, and reducing pollution through NbT technologies, such as sustainable urban drainage systems, green infrastructure, green roofs, and walls (Basu et al. 2021; Collier and Bourke 2020). Overall, these NbT technologies demonstrated positive outcomes, including benefits in food production, energy savings, rainwater storage, facade vegetation, thermal protection, biodiversity conservation, restoration, and various other mitigation strategies.

Smart Based Technology (SbT)

The latest emerging smart-based technologies (SbTs) include the Internet of Things (IoT), Artificial Intelligence (AI), and sensor technology. IoT connects the Internet and physical objects and is critical in enhancing climate resilience (Argyroudis et al. 2022). AI is an adaptable and super powerful intelligence that can assess the condition of buildings, roads, and pipeline infrastructure and predict future climate resistance.

One application of this technology is CitiCAP, an intelligent city management app launched in Lahti as part of the CitiCAP project in 2018 to support the Sustainable Urban Mobility Plan (CDP Worldwide 2024). The fundamental goal was to minimize the emissions from transport vehicles while collecting data on their mobility patterns. This data can further support the development of new transportation technologies.

Furthermore, the app visualizes each individual's carbon footprint and introduces a virtual credit currency system for procuring consumable services. Equally noteworthy is Helsingor's innovative energy-renovation model (CDP Worldwide 2024). This model, which significantly reduces buildings' energy consumption and environmental impact, was piloted in 2020 with 200 homes, resulting in an average reduction of approximately 8400 kWh and 2.7 tons of CO_2 per home. The potential for this initiative to spark a global revolution in energy efficiency is immense if scaled up.

Another key idea is to develop an intelligent area inspired by smart-based technologies established in Orebro City (CDP Worldwide 2024). The Tamarinden project (2025–2030) was developed to create an intelligent district with 700 new homes distributed across five neighbouring areas. This initiative involves innovative building designs, cutting-edge technologies, ecological motion, green farming practices, and energy conservation methods. The main aim of this project was to minimize the carbon footprint of the entire residential area. Similarly, the Smart City Control Room uses smart-based technologies to enhance public health by establishing innovative solutions for delivering smart city services. This technology facilitates the real-time monitoring of traffic, water and electricity usage, air quality, waste management, restricted traffic zones, and parking spaces. The collected data will be analyzed to inform the development of more innovative and sustainable cities. This technology was first introduced in Florence, Europe (CDP Worldwide 2024). The "The Living Lab: Sesvete" project in Zagreb City encompasses bioclimate building principles, sustainable urban planning, and strengthening urban resilience (proGIreg project 2020).

Rotterdam City in Europe has taken a proactive approach to climate resilience by developing a building with waterproofing technology as part of the "Deltaplan Tidal Park programme" (Oppla 2023). This initiative, which aims to achieve 100% waterproofing by 2025, focuses on adaptive measures such as collecting rainfall and delaying drainage. Another initiative is the Smart Control of Climate Resilience in European Coastal Cities (SCORE), which connects ten coastal cities to prevent sudden climatic disasters by providing early warning through coastal cities by coastal city living labs (SCORE 2021). However, advanced environmental sensors in coastal cities monitor pollution levels and heat waves. These technologies can be implemented globally to bolster climate resilience. By embracing these innovations, nations can tackle climate and environmental challenges more effectively.

Ecosystem Restoration and Energy Renovation

Europe is at the forefront of implementing strategies to enhance climate resilience through ecosystem restoration and energy renovation (Simonson et al. 2021). Recent studies have shown that Precision Agriculture Technologies significantly reduces environmental impact, optimizes resource utilization, and improves crop yields. Precision Agriculture is an emerging technology that is used to improve crop production with technology sensors and analysis tools (Singh et al. 2020). Adinolfi et al. (2024) discussed these technologies, which Climate Smart Agricultural Tools supplement. These tools enhance agricultural production and mitigate greenhouse gas emissions, helping farmers adjust to changing climatic conditions.

The Main River Basin project is a significant effort that employs a Systems Innovation Approach (SIA) to address climate challenges. This approach prioritizes understanding interconnected issues, such as agriculture, forestry, water, and energy management, and developing multi-sectoral strategies to tackle them. At the heart of this initiative are Living Labs, which foster collaborative creation among stakeholders and support participatory research in line with the Sustainable Development Goals (SDGs). The resources provided by SIA facilitate workshops that inspire cooperation and creative problem-solving to address future climate challenges.

The ARSINOE project, funded by the EU, combines these components and concentrates on holistic approaches for regions resilient to climate change (Rajesh et al. 2024). Within this structure, examples like the Multi-Hazard Management Tool and Strategic Resilience strategies assist stakeholders in recognizing climate risks and creating customized adaptation plans. The Climate Change Hot-Spot Identification Tool and similar tools assess areas by considering their exposure, sensitivity, and ability to adapt, which helps tailor specific adaptation strategies. These initiatives are based on a collaborative approach that engages local communities and stakeholders in the planning phase. Additionally, the research involves creating flexible adaptation plans, evaluating potential solutions under different scenarios, and conducting test runs in selected case studies (D'Oca et al. 2018).

Smart Water Collection Systems

European smart water systems are crucial in preserving water resources and tackling water shortages caused by climate change in urban environments. These systems combine energy, transportation, and urban infrastructure to improve the efficiency and management of water collection. Advanced technologies, such as intelligent sensors, Internet of Things (IoT) devices, and data analysis, are essential for enhancing water collection efficiency. These technologies make it possible to monitor water consumption, detect leaks, and optimize the allocation of resources in real time.

Decentralized stormwater management and reusing greywater from household sources such as sinks and showers enhance urban resilience to climate change. Critical in promoting these practices, public engagement tools ensure communities actively participate in water conservation efforts (Doric et al. 2024).

The SCORE project illustrates the need for creative solutions to strengthen climate resilience, particularly by implementing intelligent water collection systems. Coastal City Living Labs (CCLLs) play a crucial role in this initiative, engaging various stakeholders in the development and execution of water management systems. Furthermore, the project highlights the importance of ecosystem-based approaches (EBA), such as natural water conservation measures, which enhance ecosystem well-being. The methodologies and technologies created through the SCORE project are intended to apply to diverse coastal cities in Europe, facilitating broad adaptation to water challenges related to climate change (Figueiredo et al. 2021; Paranunzio et al. 2023).

Agricultural Resilient Practices

The SURE Act, which stands for Sustainable and Resilient EU Farming Systems, is an initiative within Europe that aims to support various sustainable farming practices to improve biodiversity and the health of ecosystems. This initiative promotes agroecology, organic farming, precision agriculture, and remote sensing as essential for developing a solid and adaptable agricultural system. Agroecological strategies, including crop diversification, conservation tillage, cover cropping, and integrated pest management, significantly encourage sustainable land use. Water management methods are also integrated to ensure the effective utilization of resources (Singh et al. 2022; Miranda et al. 2019).

Restoring ecosystems in agriculture plays a crucial role in the SURE Act, focusing on promoting biodiversity, enhancing the adaptive capacity of local stakeholders, and fostering the transformative capacity to address environmental challenges. These initiatives are vital for the restoration of ecosystems and the assurance of food security. The Act also highlights the significance of collaborating with stakeholders and promoting knowledge exchange and innovation to bolster the resilience of agricultural systems (Rajesh et al. 2024).

The European Union's agricultural research and innovation follow a strategic long-term approach focused on deriving value from land by promoting sustainable primary production and rural innovation. This approach is crucial in addressing challenges like climate change, environmental degradation, and biodiversity loss. In 2021, the European Commission highlighted soil health restoration as a significant challenge across the EU. To address this, more than 1000 local testing sites have been set up to facilitate practical experimentation and collaborative development of solutions by farmers, foresters, and urban gardeners. These sites function as showcases for effective practices that can be expanded through the Common Agricultural Policy (CAP) and other EU and national funding mechanisms (Ambros and Granvik 2020; Costantini et al. 2020; Serebrennikov et al. 2020).

Innovative Approaches to Climate Resilience

Policy and Governance Innovations

The European Commission introduced the implementation plan with the aim of the Climate Change Adaptation Mission, which supports around 150 regions and communities by 2030. The main objectives of this mission are preparation and planning, accelerating, and demonstrating systematic transformations to climate resilience (Smyth et al. 2024). This mission mainly focused on critical communities, including ecosystems and nature-based solutions, land use and food systems, water management, local economic systems, and critical infrastructure. These components are significantly impacted by climate change. Other mission establishments include the Climate-ADAPT portal, which is the primary repository on resilience and adaptation in Europe; the European Partnership for Rescuing Biodiversity for monitoring and implementing biodiversity activities; and the Research and Innovation Partnership on Climate Change and Sustainable Energy (CCSE), which helps in exchanging research and innovative ideas across countries (Smyth et al. 2024).

Community Based Adaptations (CbA)

Community-based adaptation (CbA) programs help improve the livelihoods of rural and vulnerable regions against climate change threats. CbA has been classified into many components: ecosystem integrity, resource management, microfinance and insurance, infrastructure, diversification, and building capacity (Shammin et al. 2022). It has bottom-top approaches that give hope to an uncertain world. It offers opportunities for innovation in community-engaged, resource-efficient, cost-effective, and sustainable responses to climate risks.

An Insight on Climate Resilience and Neutrality in Urban Europe

Globally, renewable energy sources have become vital for promoting climate mitigation and ensuring energy security. Cities must take advantage of this energy transformation, but how they will formulate their action plans is unclear. With the help of the European 100 Climate-Neutral and Smart Communities Mission, a new dataset was gathered that provides exclusive information on 362 cities that have stated their goal to become carbon neutral by 2030. Their insights include their degree of readiness, aspiration, capability, and the risks they see in the pursuit of zero-emission and more environmentally friendly futures (Ulpiani et al. 2023). This study primarily examines the function of renewable energy in cities' high-GHG-emitting sectors, such as buildings, transportation, waste management, and industry. It examines:

1. The current state of renewable energy production, consumption, and policy-making.
2. The significant steps to improve and scale up the deployment of renewable energy shortly.
3. How relevant instruments and policies will change over time to reduce emissions and speed up the energy transition.

Thus, for cities to implement significant renewable projects, more work must be done, obstacles must be removed, and multi-governance strategies must be put into practice.

Changeover, to demonstrate how renewable energy contributes to the goal of becoming carbon neutral. These days, the share of RES (renewable energy source) sources in cities' energy production and consumption seldom exceeds 20%. Most towns will use this as a base to encourage the use of solar energy, explore the possibilities of other renewable energy sources (including wind and geothermal), and test out circular energy models. Rather than following the conventional approach of incorporating renewable energy sources into individual buildings, cities are proposing to leverage communal systems at the community or district level to achieve improved grid stability and benefit from social, digital, and economic co-benefits. Future energy policies, however, have not included energy storage systems properly. Fascinatingly, cities emphasize the significance of RES initiatives that inter-high-emitting industries include wastewater treatment, maritime shipping, and private transportation, but these still need to be acknowledged in policy. Table 12.1 shows various factors NET Zero Initiative.

Creating, this is anticipated to keep concentrating disproportionately on public structures. Cities have recognized initiatives that could be expanded in the future and are investigating new business models, collaborations, technology, legal and regulatory frameworks, and governance structures. However, to achieve the challenging aim of climate neutrality, more work must be done, obstacles must be removed, and multi-governance strategies must be implemented (EC 2022). Figure 12.3 shows green house gases and net zero initiative factors.

Table 12.1 Various factors NET zero initiative

Green house gas release (direct)	Green house gas release (indirect)	Out of boundary green house gas release
Within city area from stationary source, transport, waste/wastewater generation and treatment etc.	Within the city boundary due to consumption of energy supply etc.	Because of the dumping and treatment of waste/wastewater generation within city area

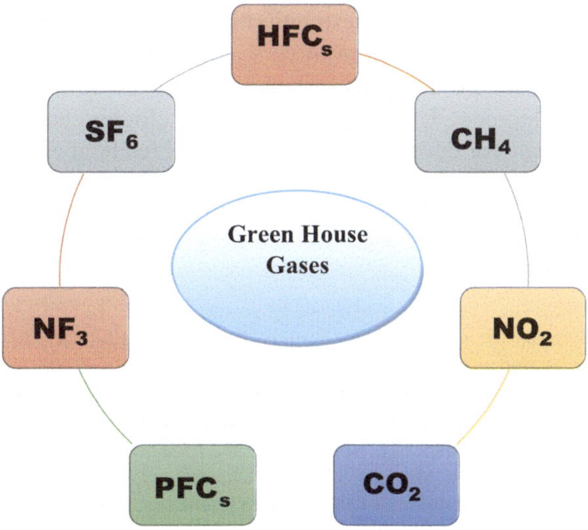

Fig. 12.3 Green house gases and NET zero initiative factors

Future Directions and Recommendations

Employing innovative solutions through science and technology in climate change is essential for constructing a resilient environment. Future recommendations from this study are as follows to help overcome the challenges of reaching a sustainable environment. A key characteristic is the expansion of international cooperation to exchange knowledge, technologies, and best practices worldwide. More attention should be paid to educating and creating awareness of the transformation to climate resilience by indicating environmental, economic, and social benefits. Create a public allowance through tax breaks and subsidies to support climate-resilient implements. New innovative solutions should come up through more advanced studies and research findings. Together, these measures contribute to developing a resilient, efficient, and inclusive future climate-resilient and sustainable environment.

Conclusion

Due to increased urbanization and climatic change, the depletion of natural resources is expected to grow. The European government has developed and established enduring solutions to these climatic challenges. This chapter examines the strategies and technologies adopted across Europe to alleviate the effects of climate change. These include the development of NbT, SbT, CbA ecosystem restoration initiatives, and European Union policies. NbT solutions focus on green infrastructure, such as buildings, rooftops, walls, and pocket parks. SbT solutions emphasize bright areas, innovative city development, and AI and IoT-based carbon footprint detection systems. Ecosystem restoration involves advancement in smart agriculture, energy conservation, and innovative water management. Furthermore, European governments have established various policies to achieve sustainable and resilient European development. Existing European technologies can be implemented globally, paving the way for developing sustainable and resilient environments.

References

Adinolfi M, Reder A, Mercogliano P, Hoy A, Milelli M, Biondi R et al (2024) Building up a digital academy in AGORA project to aware citizens, improve access to and use of climate data supporting adaptation. Copernicus Meetings No. EGU24-19094

Ambros P, Granvik M (2020) Trends in agricultural land in EU countries of the Baltic Sea region from the perspective of resilience and food security. Sustainability 12(14):5851

Argyroudis SA, Mitoulis SA, Chatzi E, Baker JW, Brilakis I, Gkoumas K, Linkov I et al (2022) Digital technologies can enhance climate resilience of critical infrastructure. Clim Risk Manag 35:100387

Arlati A, Rödl A, Kanjaria-Christian S, Knieling J et al (2021) Stakeholder participation in the planning and design of nature-based solutions. Insights from CLEVER cities project in Hamburg. Sustainability 13(5):2572

Balogun AL, Shekhar H, Budryte P, Okunola OH, Abdul-Kareem T, Abubakar IR, Yekeen ST et al (2022) Urban resilience in the fourth industrial revolution: transformative digitalization in European smart cities to address climate change. In: Resilient smart cities: theoretical and empirical insights, pp 355–373

Basu AS, Pilla F, Sannigrahi S, Gengembre R, Guilland A, Basu B et al (2021) Theoretical framework to assess green roof performance in mitigating urban flooding as a potential nature-based solution. Sustainability 13(23):13231

Bona S, Silva-Afonso A, Gomes R, Matos R, Rodrigues F et al (2022) Nature-based solutions in urban areas: a European analysis. Appl Sci 13(1):168

Calabrese M, Iandolo F, Vito P, D'Amore R, Caputo F et al (2023) Nature-based solutions: a smart way to improve urban resilience in the face of climate change. In: ITM Web of Conferences, vol 51. EDP Sciences, p 03003

Calliari E, Castellari S, Davis M, Linnerooth-Bayer J, Martin J, Mysiak J, Zandersen M et al (2022) Building climate resilience through nature-based solutions in Europe: a review of enabling knowledge, finance and governance frameworks. Clim Risk Manag 37:100450

Carvalho PN, Finger DC, Masi F, Cipolletta G, Oral HV, Tóth A, Exposito A et al (2022) Nature-based solutions addressing the water-energy-food nexus: review of theoretical concepts and urban case studies. J Clean Prod 338:130652

CDP Worldwide (2024) www.cdp.net. Accessed 28 Aug 2024

Clever Cities Project (2024) https://clevercities.eu/. Accessed 5 Sept 2024

Collier MJ, Bourke M (2020) The case for mainstreaming nature-based solutions into integrated catchment management in Ireland. Biol Environ Proc R Irish Acad 120(2):107–113

Costantini EAC, Antichi D, Almagro M, Hedlund K, Sarno G, Virto I et al (2020) Local adaptation strategies to increase or maintain soil organic carbon content under arable farming in Europe: inspirational ideas for setting operational groups within the European innovation partnership. J Rural Stud 79:102–115

D'Oca S, Ferrante A, Ferrer C, Pernetti R, Gralka A, Sebastian R, Veld P et al (2018) Technical, financial, and social barriers and challenges in deep building renovation: Integration of lessons learned from the H2020 cluster projects. Buildings 8(12):174

Đorić VS, Blagojević B, Milanović D et al (2024) Innovative approaches to water management in the smart cities of Europe. In: Zbornik radova sa Nacionalne konferencije sa medunarodnim ucešcem–Zelena Gradnja 2024, pp 341–346
Dumitru A, Wendling L (2021) Evaluating the impact of nature-based solutions: appendix of methods. Publications Office of the European Union. https://data.europa.eu/doi/10.2777/11361
Dymek D, Wilkaniec A, Bednorz L, Szczepańska M et al (2021) Significance of allotment gardens in urban green space systems and their classification for spatial planning purposes: a case study of Poznań, Poland. Sustainability 13(19):11044
European Commission, Directorate-General for Research and Innovation (2018) A new horizon for Europe: impact assessment of the 9th EU framework programme for research and innovation. Publications Office. https://data.europa.eu/doi/10.2777/194210
European Commission (2022) COM/2022/230 final—communication from the commission to the European parliament, the European Council, the Council, the European economic and social committee and the committee of the regions—REPowerEU plan. https://eur-lex.europa.eu/legal-content/EN/TXT/?uri=COM%3A2022%3A230%3AFIN. Accessed 13 Sept 2024
European Environment Agency (2024) https://www.eea.europa.eu/. Accessed 10 Sept 2024
Figueiredo I, Esteves P, Cabrita P et al (2021) Water wise—a digital water solution for smart cities and water management entities. Procedia Comput Sci 181:897–904
Hoegh-Guldberg O, Bindi M (2018) Impacts of 1.5 °C global warming on natural and human systems. 2. Notes, p 41
Kiss B, Sekulova F, Hörschelmann K, Salk CF, Takahashi W, Wamsler C et al (2022) Citizen participation in the governance of nature-based solutions. Environ Policy Gov 32(3):247–272
Koppelaar R, Marvuglia A, Havinga L, Brajković J, Rugani B et al (2021) Is agent-based simulation a valid tool for studying the impact of nature-based solutions on local economy? A case study of four European cities. Sustainability 13(13):7466
Kumar V, Ranjan D, Verma K et al (2021) Global climate change: the loop between cause and impact. In: Global climate change, pp 187–211
Mabon L, Barkved L, de Bruin K, Shih WY et al (2022) Whose knowledge counts in nature-based solutions? Understanding epistemic justice for nature-based solutions through a multi-city comparison across Europe and Asia. Environ Sci Policy 136:652–664
Mahmoud IH, Morello E, Ludlow D, Salvia G et al (2021) Co-creation pathways to inform shared governance of urban living labs in practice: lessons from three European projects. Front Sustain Cities 3:690458
Marvuglia A, Koppelaar R, Rugani B et al (2020) The effect of green roofs on the reduction of mortality due to heatwaves: results from the application of a spatial microsimulation model to four European cities. Ecol Model 438:109351
Miranda Meuwissen PM, Julie U, Mauro V, Katarzyna Z, Hugo H, Phillipa ND, Erwin W et al (2019) A framework to assess the resilience of farming systems. Agric Syst 176:102656
O'Keeffe J, Pluchinotta I, De Stercke S, Hinson C, Puchol-Salort P, Mijic A, Collins AM et al (2022) Evaluating natural capital performance of urban development through system dynamics: a case study from London. Sci Total Environ 824:153673
Oppla (2023) https://oppla.eu/case-study-finder. Accessed 5 Sept 2024
Paranunzio R, Anton I, Adirosi E, Ahmed T, Baldini L, Brandini C, Gharbia S et al (2023) A new approach towards a user-driven coastal climate service to enhance climate resilience in European cities. Sustainability 16(1):335
proGIreg project (2020) https://progireg.eu/. Accessed 5 Sept 2024
Rajesh CM, Jadhav A, Manohar KN, Bhat PP, Prasad R, Anil K, Pavan V et al (2024) A review on adaptive strategies for climate resilience in agricultural extension services in India. Arch Curr Res Int 24(6):140–150
Scharf B, Kogler M, Kraus F, Garcia Perez I, Gutierrez Garcia L et al (2021) NBS impact evaluation with GREENPASS methodology shown by the case study 'FischbekerHöfe' in Hamburg/Germany. Sustainability 13(16):9167
SCORE (2021) https://score-eu-project.eu/. Accessed 28 Aug 2024

Serebrennikov D, Thorne F, Kallas Z, McCarthy SN et al (2020) Factors influencing adoption of sustainable farming practices in Europe: a systemic review of empirical literature. Sustainability 12(22):9719

Shammin MR, Haque AE, Faisal IM et al (2022) A framework for climate resilient community-based adaptation. In: Climate change and community resilience, pp 11–30

Simonson WD, Miller E, Jones A, García-Rangel S, Thornton H, McOwen C et al (2021) Enhancing climate change resilience of ecological restoration—a framework for action. Perspect Ecol Conserv 19(3):300–310

Simperler L, Ertl T, Matzinger A et al (2020) Spatial compatibility of implementing nature-based solutions for reducing urban heat islands and stormwater pollution. Sustainability 12(15):5967

Singh P, Pandey PC, Petropoulos GP, Pavlides A, Srivastava PK, Koutsias N, Bao Y et al (2020) Hyperspectral remote sensing in precision agriculture: present status, challenges, and future trends. In: Hyperspectral remote sensing. Elsevier, pp 121–146

Singh S, Singh P, Sharma A, Choudhury M (eds) (2022) Agriculture waste management and bioresource: the circular economy perspective. Wiley. https://onlinelibrary.wiley.com/doi/book/10.1002/9781119808428

Smyth F, Murray J, Conti X, Mayor B, Madigan L, Harold T, Davies W, Dworak T, Salmoral G, Ploder M, Smithers R et al (2024) EU missions, adaptation to climate change. Activity report: May 2024

Song Y, Kirkwood N, Maksimović Č, Zheng X, O'Connor D, Jin Y, Hou D et al (2019) Nature based solutions for contaminated land remediation and brownfield redevelopment in cities: a review. Sci Total Environ 663:568–579

Ulpiani G, Vetters N, Shtjefni D, Kakoulaki G, Taylor N (2023) Let's hear it from the cities: on the role of renewable energy in reaching climate neutrality in urban Europe. Renew Sustain Energy Rev 183:113444. https://doi.org/10.1016/j.rser.2023.113444

United Nations Environment Programme (2020) Global status report for buildings and construction: towards a zero-emission, efficient and resilient buildings and construction sector. Nairobi

Zwierzchowska I, Haase D, Dushkova D et al (2021) Discovering the environmental potential of multi-family residential areas for nature-based solutions. A Central European cities perspective. Landsc Urb Plan 206:103975

Part IV
Ecosystems in the Climate Resilience Equation: Measuring, Predicting, and Shaping the Future

Chapter 13
A Significance of Climate Resilience: Strategies for Local, Regional, and National Levels

T. R. Kumaraswamy, Moharana Choudhury, Shivani Ponappa, and Sushobhan Majumdar

Abstract This chapter delves into the dynamic nature of resilience across different scales—local, regional, and national. Resilience is the capacity to anticipate, prepare for, respond to, and recover from adverse events, is crucial for maintaining stability and adaptability in various challenges. This chapter begins by outlining the conceptual foundations of resilience and its significance at different governance levels. It then presents a detailed examination of resilience strategies at each scale, highlighting local community-based initiatives, regional coordination efforts, and national policy frameworks. The chapter evaluates the effectiveness of these strategies, identifies familiar challenges, and suggests practical recommendations for enhancing resilience. Local resilience is shown through grassroots efforts and government policies, while regional resilience emphasizes collaborative planning and infrastructure development. The focus shifts to federal policies and large-scale infrastructure investments at the national level to bolster national stability and adaptability. A key theme of the chapter is the need for an integrated approach that harmonizes resilience efforts across all levels. The discussion underscores the benefits of coordinated strategies in managing risks and building a robust, adaptive society. The chapter concludes with insights into future research directions and the importance of cross-level collaboration in advancing resilience.

T. R. Kumaraswamy
Centre for Sustainability, School of Law, Governance and Public Policy, Chanakya University, Bengaluru, India

M. Choudhury
Environmental Research and Management Division, Assam, Voice of Environment (VoE) Guwahati, Guwahati, India

S. Ponappa
School of Biotechnology, University of Leads, Leads, UK

S. Majumdar (✉)
Department of Geography, Pandit Raghunath Murmu Smriti Mahavidyalaya, Bankura, West Bengal, India
e-mail: sushobhan91@gmail.com

M. Choudhury et al. (eds.), *Sustainable Synergy: Harnessing Ecosystems for Climate Resilience*, Climate Change Management,
https://doi.org/10.1007/978-3-031-77957-2_13

Keywords Resilience · Local strategies · Regional coordination · Integrated approaches · Climate change

Introduction

In the present era, society is growing increasingly concerned about the substantial repercussions of climate change and expects to encounter similar challenges in the future. The rising atmospheric and oceanic temperatures provide evidence of significant global warming in recent years (Haldar et al. 2023). The rise in temperature has put biodiversity at risk and undermined ecosystems (ISDR 2009). The substantial increase in global greenhouse gas (GHG) emissions is the primary cause of climate change. Since the Industrial Revolution, the concentrations of carbon dioxide, methane, and other greenhouse gases in the atmosphere have nearly doubled, resulting in numerous environmental issues (Mondal et al. 2023). Human activities, such as burning fossil fuels, changing how land is used, decomposing waste in landfills, and extracting natural gas, have significantly increased the release of these emissions (IPCC 2007a, b; Rajpal et al. 2024). Based on future projections, emissions are expected to continue rising, especially in developing nations. Developing nations will face substantial challenges, including insufficient food supply and limited access to water, affecting billions of people (UNFCCC 2007). Developing nations are highly vulnerable to the adverse impacts of climate change. Currently, the country is dealing with the depletion of water resources, intense heat waves, prolonged droughts, storms, and floods, all of which pose more significant risks to people's well-being and ability to support them (GCRI et al. 2009). Advancement in this field depends on the critical establishment and execution of state-of-the-art governance systems. Enhancing the organizational capacity of institutions is crucial for effectively implementing CCA measures, thereby playing a vital function in the success of both CCA and CCM endeavours. To bolster the resilience of communities and nations against climate change, it is imperative to establish effective systems of governance and mechanisms that promote accountability and openness in both public and private institutions. This chapter explores the latest techniques for improving the resilience of local, regional, and national areas. The text examines potential paths for incorporating hierarchical and participatory methods, tactics for enhancing capability, and the necessary governance systems.

The International Framework for Climate Resilience is a global initiative to enhance countries' ability to acclimatize and withstand the outcomes of climate change. The focus on climate resilience has shifted from providing immediate relief and response to prioritizing longstanding efforts to reduce the impact of climate change and be prepared for its effects. Multiple global frameworks have recognized the imperative of sustainable development to demonstrate resilience in catastrophes and acclimatize to climate change (NDMA 2019; Gupta et al. 2016).

The Paris Agreement, established under the United Nations Framework Convention on Climate Change (UNFCCC), is a legally enforceable to tackle climate change by promoting sustainable practices and investments that decrease GHG emissions and lower carbon footprints (UNFCCC 2011). The agreement received official approval on the 12th of December 2015 at the 21st Conference of Parties (COP 21), following thorough deliberation by 196 participants. The results of COP21 established a distinct correlation between promoting socio-economic progress, preserving the environment, and executing sustainable land management tactics. The conference suggested implementing a proactive strategy to effectively reduce and reverse the desertification and land degradation process while simultaneously responding to the consequences of climate change in the affected regions. The establishment of Sustainable Development Goals (SDGs) has been facilitated by international cooperation and progress in climate policy.

The connection between adaptation and resilience to climate change is intricately linked and can be applied across various sectors and facets of growth. Ten of the seventeen SDGs are directly or indirectly associated with climate resilience. The SDG framework comprises 25 targets that emphasize the significance of developing the competence of communities and nations to achieve the SDGs while also addressing the pressing need to minimize the impacts of climate change. The goals of tackling poverty, hunger, well-being, infrastructure, education, water management, climate change, sustainable urban areas, and marine and terrestrial ecosystems explicitly incorporate the principles of adapting to and building resilience against climate change. Figure 13.1 shows integration of resilience strategies across governance strategy.

Fig. 13.1 Integration of resilience strategies across governance strategy

Conceptual Framework

Theoretical foundations of resilience emphasize the ability of systems to absorb disturbances, adapt, and transform in the face of challenges. Resilience theory, rooted in ecology and systems science, views resilience as a multifaceted capacity involving robustness, adaptability, and transformability (Holling 1973; Gunderson and Holling 2002). Robustness refers to a system's capacity to endure shocks without undergoing significant changes, while adaptability denotes the system's ability to adjust strategies in response to evolving conditions (Folke et al. 2010). Transformability involves the capacity to change and reorganize fundamentally in response to extreme stressors (Walker et al. 2004). Recent research has expanded these concepts to include the social dimensions of resilience, integrating socio-ecological systems approaches highlighting the interplay between human and environmental factors (Cox and Perry 2019). This perspective underscores the importance of social networks, governance structures, and community engagement in building resilience across different scales. Figure 13.2 shows path to sustainability from local, regional to national level.

Local, regional, and national resilience each address distinct scales of governance and response but are interconnected through their collective impact on overall system stability. Local resilience focuses on community-level capacities to manage and

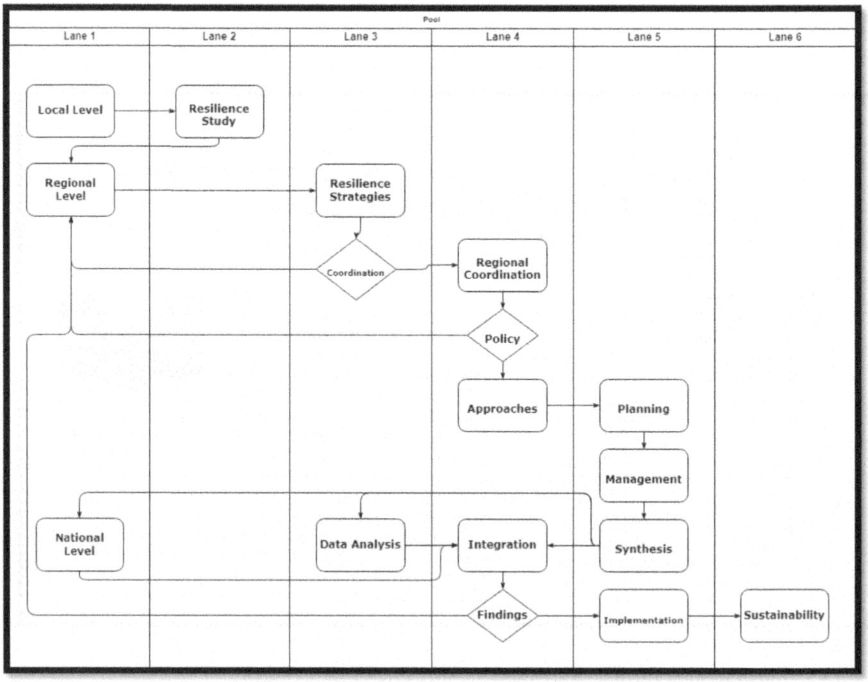

Fig. 13.2 Path to sustainability from local, regional to national level

recover from disruptions, emphasizing grassroots initiatives, local governance, and social networks (McAslan 2014). This scale is crucial for immediate responses and adaptive practices tailored to local needs. Regional resilience operates at a broader scale, involving coordination among multiple localities within a larger geographic area. It highlights the importance of regional planning, infrastructure, and inter-local collaborations to address challenges that span multiple communities (Cox and Perry 2019). Regional resilience often involves economic integration and shared resources, requiring cohesive strategies across jurisdictions.

National resilience encompasses policies and frameworks to enhance a country's capacity to withstand and recover from widespread crises. It includes federal regulations, national infrastructure investments, and macroeconomic strategies (Holling et al. 2015). National resilience provides the overarching framework within which local and regional resilience efforts operate, ensuring consistency and support at broader scales. These scales are connected through mutual dependency: local and regional resilience feeds into national stability, while national policies support and enhance localized and regional efforts.

Several vital factors significantly influence resilience across scales, including social capital, infrastructure, and governance. Social capital encompasses the networks, norms, and trust that facilitating community coordination and cooperation (Putnam 2000). High social capital enhances a community's ability to mobilize resources, support recovery efforts, and adapt to changes effectively (Woolcock and Narayan 2000). Infrastructure plays a critical role in resilience by providing the physical and technological systems necessary for functionality and recovery. Resilient infrastructure—such as transportation networks, utilities, and communication systems—enables quicker response and recovery during disruptions (Rose 2004). Robust, adaptable infrastructure investments are crucial for maintaining service continuity and reducing hazard vulnerability. Governance involves the structures, policies, and processes that guide decision-making and resource allocation. Effective governance ensures that resilience strategies are well-coordinated, equitable, and responsive to emerging risks (Pelling 2011). Robust governance frameworks facilitate cross-sector collaboration, enforce regulations, and integrate resilience considerations into planning and policymaking.

Local Resilience Strategies

Local resilience refers to communities' and institutions' capacity to withstand, adapt to, and recover from disruptions such as natural disasters, economic shocks, or social upheavals. It involves a community effectively leveraging its unique social networks, local knowledge, and resources to manage and rebound from adverse events (Béné et al. 2022). The importance of local resilience lies in its direct impact on immediate recovery and adaptation efforts (Fig. 13.2). Local communities are often the first responders to crises and play a crucial role in mitigating the effects of disruptions through localized responses and support systems (Arup 2022). Strengthening local resilience ensures that communities can maintain essential functions, reduce vulnerability, and enhance their ability to recover independently of external aid. Local resilience also contributes to broader stability by fostering adaptive capacities and self-sufficiency, critical for reducing the overall burden on regional and national systems (Mastrorillo et al. 2021). Effective local resilience strategies can lead to more sustainable and equitable recovery processes, making them vital for long-term community well-being and stability.

Evaluating the effectiveness of resilience strategies involves assessing how well they enhance a system's ability to withstand, adapt to, and recover from disturbances. Effectiveness is often measured through reduced recovery times, improved resource mobilization, and enhanced adaptive capacities (Smith and Penning-Rowsell 2021). Successful resilience strategies demonstrate their impact through increased community engagement, effective risk management, and sustained functionality of essential services. However, challenges persist in the implementation and evaluation of these strategies. Common issues include inadequate funding, fragmented governance structures, and insufficient integration of local knowledge (Johnson et al. 2023).

Additionally, resilience efforts may face difficulties addressing systemic inequalities, which can hinder equitable outcomes and limit overall effectiveness (Béné et al. 2022). Recent evaluations highlight the need for comprehensive approaches that consider immediate and long-term impacts ensure stakeholder inclusivity, and foster adaptive learning (Hughes et al. 2022). Addressing these challenges requires continuous monitoring, iterative improvements, and a commitment to bridging gaps between policy, practice, and community needs.

To enhance local resilience, communities should strengthen social networks and foster active community engagement, as these networks are crucial for mobilizing resources and supporting collective action during crises (Béné et al. 2022). Investing in robust and adaptable infrastructure is essential to withstand and recover from disruptions, with a focus on integrating future risks into planning and design (Arup 2022). Encouraging local knowledge and involving community members in resilience planning ensures that strategies are relevant and effectively address specific local needs (Hughes et al. 2022). Additionally, increasing risk awareness through education empowers residents to prepare and respond proactively (Smith and Penning-Rowsell 2021). Securing sustainable funding and leveraging partnerships with private and non-profit sectors can also support and sustain resilience initiatives over the long term (Johnson et al. 2023). Implementing these recommendations can build stronger, more adaptive communities capable of effectively managing and recovering from various challenges (Fig. 13.3).

Fig. 13.3 Community-based Adaptation by villagers (The restroom shed, Photo Captured by: Moharana Choudhury)

Regional Resilience Strategies

Regional resilience refers to the capacity of a geographic area, encompassing multiple communities or jurisdictions, to withstand, adapt to, and recover from various disruptions such as natural disasters, economic shifts, or social challenges. It involves coordinated planning and resource management across local entities to address region-wide risks and enhance collective response capabilities (Mastrorillo et al. 2021). The importance of regional resilience lies in its ability to address issues beyond individual communities, such as regional infrastructure, economic interdependencies, and shared environmental resources (Cox and Perry 2022). By fostering collaboration among local governments and stakeholders, regional resilience promotes more effective disaster response, sustainable development, and equitable resource distribution, thereby strengthening the overall stability and adaptability of the region (Arup 2022). Evaluating the effectiveness of resilience strategies involves assessing their impact on a community's ability to manage and recover from disruptions using metrics such as recovery speed and adaptability (Smith and Penning-Rowsell 2021). However, challenges often arise, including insufficient funding, fragmented implementation, and difficulties integrating diverse stakeholder inputs (Johnson et al. 2023). Recent evaluations highlight that while some strategies effectively enhance local and regional resilience, systemic inequalities and coordination gaps can undermine their impact (Béné et al. 2022). Addressing these challenges requires a comprehensive approach, continuous monitoring, and adaptive management to ensure strategies remain effective and equitable over time (Hughes et al. 2022). To enhance regional resilience, it is crucial to promote collaborative planning among local governments and stakeholders

Fig. 13.4 Water resource management in the high hill area (Photo Captured by: Moharana Choudhury)

to address shared risks and optimize resource use (Cox and Perry 2022). Investing in adaptive infrastructure that can withstand and quickly recover from disruptions ensures the continuity of essential regional services (Arup 2022). Strengthening regional governance structures to facilitate effective coordination and decision-making improves the integration of resilience strategies (Mastrorillo et al. 2021). Additionally, fostering community engagement and local knowledge helps tailor resilience efforts to specific regional needs and enhances local support (Béné et al. 2022). Regular monitoring and evaluating resilience initiatives allow for adjustments and improvements based on performance and emerging challenges, ensuring ongoing effectiveness (Smith and Penning-Rowsell 2021) (Fig. 13.4).

National Resilience Strategies

National resilience refers to the ability of a country to withstand, adapt to, and recover from widespread disruptions, including natural disasters, economic crises, and geopolitical conflicts. It encompasses developing and implementing comprehensive policies and frameworks that support the coordination and integration of efforts across all levels of government and society (Mastrorillo et al. 2021). The importance of national resilience lies in ensuring the stability and functionality of essential services, maintaining economic stability, and safeguarding national security. Effective national resilience strategies provide a robust foundation for supporting regional and local initiatives, enhancing overall preparedness and response capabilities (Arup 2022). By addressing systemic risks and facilitating resource allocation, national resilience plays a critical role in mitigating the impacts of large-scale disruptions and promoting long-term stability and recovery (Cox and Perry 2022). Evaluating the effectiveness of resilience strategies involves assessing their impact on a system's ability to recover from disruptions and adapt to changing conditions. Recent studies highlight that successful resilience measures are characterized by improved response times, enhanced community engagement, and better resource management (Smith and Penning-Rowsell 2021). However, significant challenges persist, such as insufficient funding, fragmented implementation, and difficulties integrating diverse stakeholder inputs (Johnson et al. 2023). Additionally, strategies often struggle with addressing systemic inequalities, undermining their effectiveness and equitable outcomes (Béné et al. 2022). To overcome these challenges, continuous monitoring, adaptive management, and a focus on inclusivity are essential for refining resilience approaches and ensuring they meet evolving needs (Hughes et al. 2022).

To enhance national resilience, policymakers should strengthen inter-agency coordination to ensure cohesive and comprehensive response efforts across all levels of government (Mastrorillo et al. 2021). Investing in robust infrastructure and integrating advanced technologies for risk management can improve preparedness and response capabilities (Arup 2022). Additionally, promoting inclusive policies that address systemic inequalities ensures that resilience measures benefit all populations equitably (Béné et al. 2022). Regular evaluation and adaptive management of resilience strategies are crucial for maintaining effectiveness in the face of evolving risks (Smith and Penning-Rowsell 2021).

Integrative Approaches to Resilience

A coordinated approach is crucial for effective resilience planning and response across local, regional, and national levels. Such coordination ensures that strategies are aligned, resources are optimized, and efforts are mutually reinforcing, enhancing overall effectiveness (Mastrorillo et al. 2021). It facilitates consistent communication, integrated risk management, and comprehensive support, addressing risks that span multiple jurisdictions (Arup 2022). This alignment also helps avoid duplication of efforts and gaps in coverage, ultimately leading to a more robust and adaptable resilience framework (Smith and Penning-Rowsell 2021). To integrate local, regional, and national resilience efforts, it is essential to establish clear communication channels and shared goals across all levels (Mastrorillo et al. 2021). Implementing coordinated planning frameworks that align policies and actions ensures consistency and efficiency (Arup 2022). Fostering collaborative partnerships among stakeholders at different levels helps pool resources and expertise (Cox and Perry 2022). Regular cross-level meetings and joint exercises can enhance mutual understanding and readiness, while data sharing and transparency facilitate informed decision-making and unified responses (Smith and Penning-Rowsell 2021).

Challenges in coordinating resilience efforts include fragmented authority and inconsistent priorities across different levels of government (Mastrorillo et al. 2021). These issues can lead to overlapping efforts and gaps in coverage. Solutions involve establishing clear governance structures and developing unified frameworks that align objectives and streamline processes (Arup 2022). Enhanced communication protocols and regular joint exercises can improve coordination and ensure that strategies are synchronized effectively (Cox and Perry 2022). Additionally, fostering collaborative networks among stakeholders helps address and overcome these coordination barriers (Smith and Penning-Rowsell 2021).

Comparative Analysis

Resilience strategies vary significantly across local, regional, and national levels, reflecting their distinct scopes and challenges. Local resilience strategies focus on immediate, community-specific needs, emphasizing grassroots engagement, localized infrastructure improvements, and direct disaster response mechanisms (Arup 2022). In contrast, regional resilience involves broader coordination among multiple jurisdictions, integrating regional infrastructure and resources to address shared risks and enhance collective recovery efforts (Cox and Perry 2022). On the other hand, national resilience strategies encompass comprehensive policies and frameworks designed to support and unify lower-level efforts, ensuring resource allocation, regulatory standards, and long-term strategic planning (Mastrorillo et al. 2021). While local strategies are more tailored and immediate, regional and national strategies require higher-level coordination and broader scope to address systemic risks and provide overarching support (Smith and Penning-Rowsell 2021).

Policy Implications

Policymakers at all levels must prioritize integrated resilience planning to ensure coherence and effectiveness across local, regional, and national strategies. At the local level, fostering community engagement and investing in adaptable infrastructure is crucial for immediate and effective responses to disruptions (Arup 2022). Regional policymakers should focus on facilitating cross-jurisdictional collaboration and coordinating resources to address shared risks and vulnerabilities (Cox and Perry 2022). At the national level, the emphasis should be on creating supportive frameworks and providing financial and technical resources to empower local and regional efforts while ensuring policies address systemic risks and inequalities (Mastrorillo et al. 2021). Effective communication and regular evaluation of resilience strategies at all levels are essential for adapting to emerging challenges and ensuring long-term sustainability (Smith and Penning-Rowsell 2021).

To enhance resilience through policy development and implementation, promoting cross-sectoral integration by aligning policies across local, regional, and national levels is essential to ensure cohesive and comprehensive strategies (Smith and Penning-Rowsell 2021). Encouraging stakeholder participation is crucial; engaging communities, businesses, and non-governmental organizations in policy-making ensures practical strategies and addresses local needs (Béné et al. 2022). Investing in data-driven approaches helps assess risks and track the effectiveness of resilience measures, enabling evidence-based decision-making (Arup 2022). Policymakers should also prioritize flexibility and adaptability in resilience plans to accommodate evolving challenges and integrate lessons learned from past experiences (Cox and Perry 2022). Finally, securing adequate funding and resources is vital for successfully implementing and sustaining resilience initiatives (Mastrorillo et al. 2021)—the potential for cross-level collaborations.

Future Research Directions

Future studies on resilience strategies should evaluate the long-term impacts and effectiveness of implemented measures, including how they adapt to evolving risks and changing conditions (Smith and Penning-Rowsell 2021). Research should also explore comparative analyses of resilience strategies across different contexts and scales to identify best practices and lessons learned (Cox and Perry 2022). Additionally, investigating the role of emerging technologies in enhancing resilience, such as data analytics and early warning systems, can provide insights into innovative approaches (Arup 2022). Studies should also address equity and inclusion in resilience planning to ensure that vulnerable populations benefit from resilience measures (Béné et al. 2022). Finally, examining the integration of resilience strategies into broader policy frameworks and the effectiveness of cross-sectoral coordination could reveal ways to improve overall resilience governance (Mastrorillo et al. 2021).

Emerging trends in resilience strategies highlight the growing integration of technological innovations, such as artificial intelligence and big data, to enhance predictive capabilities and real-time decision-making (Arup 2022). There is also increasing emphasis on nature-based solutions and sustainability, aiming to incorporate ecological resilience into urban and regional planning (Cox and Perry 2022). However, future challenges include addressing systemic inequalities that may exacerbate vulnerabilities, requiring more inclusive and equitable resilience approaches (Béné et al. 2022). Additionally, climate change presents ongoing risks, necessitating adaptive strategies that can handle the increasing frequency and severity of extreme events (Mastrorillo et al. 2021). Policymakers must also navigate complex governance structures and ensure effective coordination among various stakeholders to implement comprehensive and cohesive resilience strategies (Smith and Penning-Rowsell 2021).

Conclusion

In conclusion, the effectiveness of resilience strategies across local, regional, and national levels hinges on their ability to integrate diverse approaches and address varying risk scales. Local resilience thrives on community engagement and adaptable infrastructure, while regional resilience benefits from coordinated planning and cross-jurisdictional collaboration. At the national level, comprehensive policies and resource allocation underpin overall stability and support lower-level efforts. Emerging trends such as technological advancements and nature-based solutions offer promising avenues for enhancing resilience, but challenges like systemic inequalities and climate change require ongoing attention. Effective resilience strategies must continuously evolve, incorporating lessons learned and adapting to new risks. Policymakers and stakeholders must foster inclusive and flexible approaches to build robust, adaptive systems that can withstand and recover from an array of disruptions, ensuring long-term sustainability and stability across all levels of governance.

References

Arup (2022) Resilient cities: strategies for urban resilience. Arup Publishing

Béné C, Newsham A, Davies M, Godfrey-Wood R (2022) Local resilience and social capital: lessons from community-based adaptation projects. Routledge

Cox RS, Perry KM (2019) Social resilience and disaster recovery: a new approach. J Environ Manage 245:284–294

Cox RS, Perry KM (2022) Social resilience and regional recovery: insights from recent challenges. J Reg Stud 56(3):430–444

Folke C, Carpenter SR, Elmqvist T, Gunderson L, Walker B (2010) Resilience thinking: sustaining ecosystems and people in a changing world. Island Press

GCRI et al (2009) Climate change and India: a 4X4 assessment. Indian Network for Climate Change Assessment (INCCA) Report

Gunderson LH, Holling CS (2002) Panarchy: understanding transformations in human and natural systems. Island Press

Gupta AK, Nair SS, Singhal P (2016) Disaster management and risk reduction: role of environmental knowledge. Indian Institute of Technology, Kanpur

Haldar S, Choudhury M, Choudhury S, Samanta P (2023) Trend analysis of long-term meteorological data of a growing metropolitan city in the era of global climate change. Total Environ Res Themes 7:100056. https://doi.org/10.1016/j.totert.2023.100056

Holling CS (1973) Resilience and stability of ecological systems. Annu Rev Ecol Syst 4:1–23

Holling CS, Gunderson LH, Peterson GD (2015) Panarchy: understanding transformations in human and natural systems. Island Press

Hughes TP, Bellwood DR, Folke C (2022) The resilience of ecosystems and the challenge of adapting to climate change. Ecol Soc 27(3):15

IPCC (2007a) Climate change 2007: the physical science basis. In: Contribution of Working Group I to the fourth assessment report of the Intergovernmental Panel on Climate Change. Cambridge University Press

IPCC (2007b) Climate change 2007: impacts, adaptation and vulnerability. In: Contribution of Working Group II to the fourth assessment report of the Intergovernmental Panel on Climate Change. Cambridge University Press
ISDR (2009) Global assessment report on disaster risk reduction. United Nations International Strategy for Disaster Reduction, Geneva, Switzerland
Johnson L, O'Neil B, Clark D (2023) Evaluating urban resilience strategies: effectiveness and implementation challenges. J Urban Aff 45(2):311–326
Mastrorillo M, Jha S, Young C (2021) Building regional resilience through coordinated adaptation strategies. Glob Environ Chang 67:102213
McAslan D (2014) Building resilience at the local level: Challenges and opportunities. Int J Disaster Risk Reduct 10:120–130
Mondal T, Choudhury M, Kundu D, Dutta D, Samanta P (2023) Landfill: an eclectic review on structure, reactions and remediation approach. Waste Manage 164:127–142. https://doi.org/10.1016/j.wasman.2023.03.034
Rajpal A, Choudhury M, Goswami S, Chakravorty A, Raghavan V (eds) (2024) Waste management and treatment: advances and innovations, 1st edn. CRC Press. https://doi.org/10.1201/9781003258377
Smith D, Penning-Rowsell E (2021) Evaluating resilience strategies: case studies and lessons learned. Springer
NDMA (2019) National disaster management plan. National Disaster Management Authority, Government of India
Pelling M (2011) Adaptation to climate change: from resilience to transformation. Routledge
Putnam RD (2000) Bowling alone: the collapse and revival of American community. Simon & Schuster
Rose A (2004) Economic resilience to disasters: toward a theory and its application. Disasters 28(1):1–18
UNFCCC (2007) Climate change: impacts, vulnerabilities and adaptation in developing countries. United Nations Framework Convention on Climate Change
UNFCCC (2011) Paris agreement. United Nations Framework Convention on Climate Change. https://unfccc.int/sites/default/files/english_paris_agreement.pdf
Walker B, Holling CS, Carpenter SR, Kinzig A (2004) Resilience, adaptability, and transformability in social-ecological systems. Ecol Soc 9(2):5
Woolcock M, Narayan D (2000) Social capital: implications for development theory, research, and policy. World Bank Res Observer 15(2):225–249

Chapter 14
Building a Resilient Tomorrow: Ecosystem-Based Approaches to Measurement, Assessment, Prediction, and Future Perspectives in Climate Resilience

Banani Ray Chowdhury, Debopriyo Pal, Saptaswa Chakraborty, Induja Samanta, Mihika Bhattacharjee, Shreya Sen, Soutrik Bhattacharya, Zacharias Frontistis, and Sushobhan Majumdar

Abstract Climate resilience is a complex and often contested concept, vulnerable to varying interpretations that can serve different interests and actors. This chapter presents a framework for analyzing and categorizing climate resilience discourses, focusing on their conceptual foundations, key targets, and institutional advocates. The Convention on Biological Diversity and the United Nations Framework Convention on Climate Change advocate for Ecosystem-based Adaptation as a crucial climate change strategy. EbA enhances resilience by leveraging biodiversity and ecosystem services. Critical practices include Restoring mangroves to mitigate storm surges. Managing watersheds are to reduce droughts and floods. Implementing rangeland management to prevent desertification sustainably managing fisheries and forests are to secure food supplies. These approaches support climate adaptation and protect and sustain vital ecosystems. These approaches shield communities

Equal contribution: Debopriyo Pal and Saptaswa Chakraborty.

B. R. Chowdhury (✉) · D. Pal · S. Chakraborty · I. Samanta · M. Bhattacharjee · S. Sen · S. Bhattacharya
Department of Biotechnology, Bengal Institute of Technology, Kolkata, West Bengal, India
e-mail: bananiraychowdhury683@gmail.com

B. R. Chowdhury
Baranagar Baghajatin Social Welfare Organisation, Kolkata, West Bengal, India

World Education & Culture Trust, Kolkata, West Bengal, India

Z. Frontistis
Department of Chemical Engineering, University of Western Macedonia, Kozani, Greece

S. Majumdar
Department of Geography, Pandit Raghunath Murmu Smriti Mahavidyalaya, Bankura, West Bengal, India

M. Choudhury et al. (eds.), *Sustainable Synergy: Harnessing Ecosystems for Climate Resilience*, Climate Change Management,
https://doi.org/10.1007/978-3-031-77957-2_14

from climate impacts, such as extreme weather events, and offer co-benefits like biodiversity conservation and improved livelihoods. Integrating EbA into policy and planning provides adaptive, cost-effective, and sustainable solutions, fostering long-term resilience for both human and natural systems. Few modern climate resilience strategies involve ecological alternatives, such as restoring coastal ecosystems for the sequestration of blue carbon and adaptive land management to enhance biodiversity. Together with advanced CO_2 capture techniques like microalgae cultivation, these strategies provide viable means to slow global warming while preserving vital ecological functions.

Keywords Carbon sequestration · Climate resilience · Biological diversity · Sustainable management

Introduction

Ecosystem-based approaches (EbA) are an essential strategy to improve climate resilience because they provide a solution based on nature that simultaneously adapts to climate change, mitigation, and preservation of biodiversity (Cohen-Shacham et al. 2016). Such approaches capitalize on ecosystems' innate ability to serve as a protective barrier against the effects of climate change, with several additional benefits for both ecological integrity and human well-being (Seddon et al. 2020). The latest study has shown how effectively EbA boosts resilience in various habitats. Restoring mangroves and seagrass meadows in coastal locations is beneficial in reducing the energy of waves, lessening the effects of storm surges, and sequestering significant amounts of blue carbon (Macreadie et al. 2021). New studies on microalgae cultivation show promise for producing biofuel and higher sequestering carbon capacity by wetlands, providing a dual benefit for mitigating climate change (Gogoi et al. 2024a, b; Mobin and Alam 2017). Urban green infrastructure programs have demonstrated promise in mitigating the influence of the heat islands in cities' effects and controlling stormwater runoff, thereby enhancing a city's ability to withstand severe weather conditions caused by variations in the climate (Keesstra et al. 2018). The crucial role that adaptive land management techniques like conservation agriculture and agroforestry play in boosting biodiversity and strengthening climate resilience is becoming more widely acknowledged (Gogoi et al. 2024a, b; Singh et al. 2022). These methods support landscape-scale conservation initiatives by enhancing soil health, retaining water, and facilitating wildlife habitat connectivity (Kremen and Merenlender 2018). The most recent study emphasizes how affordable EbA is compared to conventional grey infrastructure alternatives. According to a global review, over 3.3 billion people might benefit from natural ways to adapt to climate change, and they would be 28% less expensive than artificial alternatives (Chausson et al. 2020).

Furthermore, various environmental services, like better air and water standards, increased biodiversity, and more recreational opportunities, are frequently produced by EbA initiatives (Chausson et al. 2020). The implementation of EbA confronts obstacles despite its potential, including inadequate finance processes, a lack of policy integration, and doubts about its long-term efficacy in light of changing climate circumstances (Kabisch et al. 2016). Nonetheless, new developments in predictive modelling and monitoring technology strengthen our capacity to develop and evaluate EbA solutions (Lavorel et al. 2020). Ecosystem-based approaches provide a sustainable and comprehensive solution to improve resilience as we negotiate the difficulties of climate change (Kumar and Choudhury 2021). These initiatives, which closely correspond with global sustainability goals, address immediate climate threats while promoting long-term ecological health and human well-being by utilizing natural resources, such as coastal habitats and microalgae production (Kok et al. 2021). Figure 14.1 shows ecosystem based approach and its effects.

The chapter's objective is to investigate how ecosystem-based approaches can enhance climate resilience through comprehensive measurement, assessment, and prediction techniques. The study aims to evaluate current methodologies, explore innovative tools and models, and identify future trends in ecosystem-based resilience strategies. By integrating these approaches into policy and practice, the chapter seeks to provide actionable insights and recommendations for building robust, adaptive systems that can effectively respond to the challenges posed by climate change.

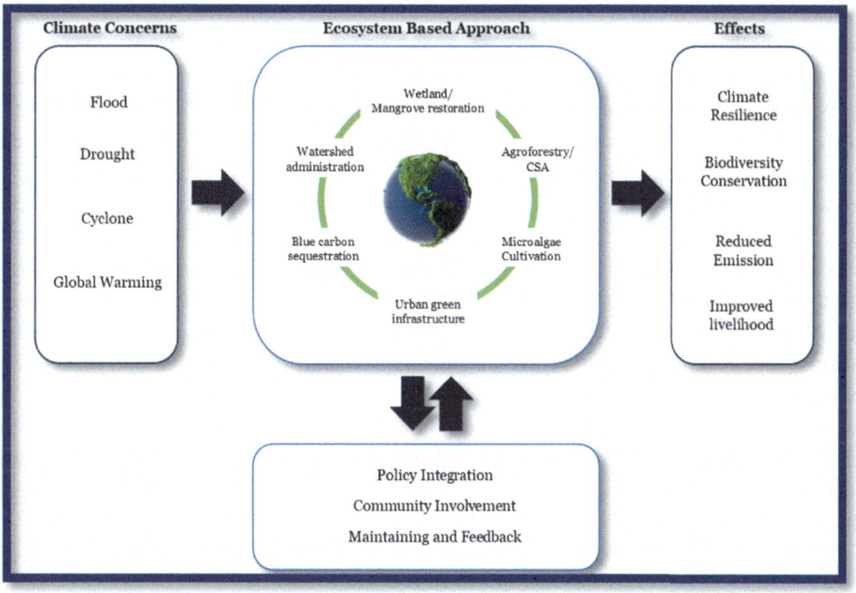

Fig. 14.1 Ecosystem based approach and its effects

Theoretical Foundations of Ecosystem-Based Approaches in Climate Resilience

The history of ecosystem-based approaches to climate resilience can be traced back to the 1970s when Holling (1973) introduced the notion of ecological resilience. This signified a transition from equilibrium-centered perspectives to those that acknowledge the dynamic nature of ecosystems. The study rapidly extended to include social-ecological systems, recognizing the complex interplay between human and natural systems (Folke 2006). Ecosystem services, resilience, and adaptability are key ideas in this discipline. Understanding resilience is essential to comprehending how ecosystems and communities endure the stress of climate change. Adaptation entails adjusting to human or natural systems to lessen vulnerability to the effects of climate change (IPCC 2014). According to (Reid et al. 2005) ecosystem services are the advantages that ecosystems offer, like food, water, and climate regulation. By incorporating ecological, social, and economic viewpoints, frameworks such as the socio-ecological systems approach are formed, emphasizing the profound connection between ecosystem health and human well-being (Ostrom 2009). Recent research has increasingly emphasized nature-based solutions (NbS) to improve climate resilience. NbS use ecosystems to address social concerns like climate change adaptation and mitigation (Seddon et al. 2020). Mangrove restoration, for example, has been shown to preserve coastal areas while sequestering carbon and supporting local livelihoods (Spalding et al. 2014).

Similarly, urban green spaces have been acknowledged for mitigating heat island effects and controlling stormwater runoff (Kabisch et al. 2016). The field is developing with an emphasis on adaptive management, stakeholder participation, and the use of indigenous and local knowledge in ecosystem-based climate resilience. Embracing ambiguity in complex systems, adaptive management encourages adaptability and ongoing learning. The integration of varied viewpoints in resilience plans is ensured through stakeholder engagement, and Indigenous knowledge provides crucial information for managing ecosystems in the face of climate change (Berkes et al. 2000; Reid et al. 2009). To optimize the co-benefits of ecosystem-based interventions, recent advancements have emphasized versatile landscapes and the synergies between mitigation and adaptation to climate change (Lavorel et al. 2020). A growing trend is toward transformative adaptation, which fundamentally modifies social and ecological systems to address the underlying causes of vulnerability and foster long-term resilience (Pelling et al. 2015). Figure 14.2 shows methodological framework of the study.

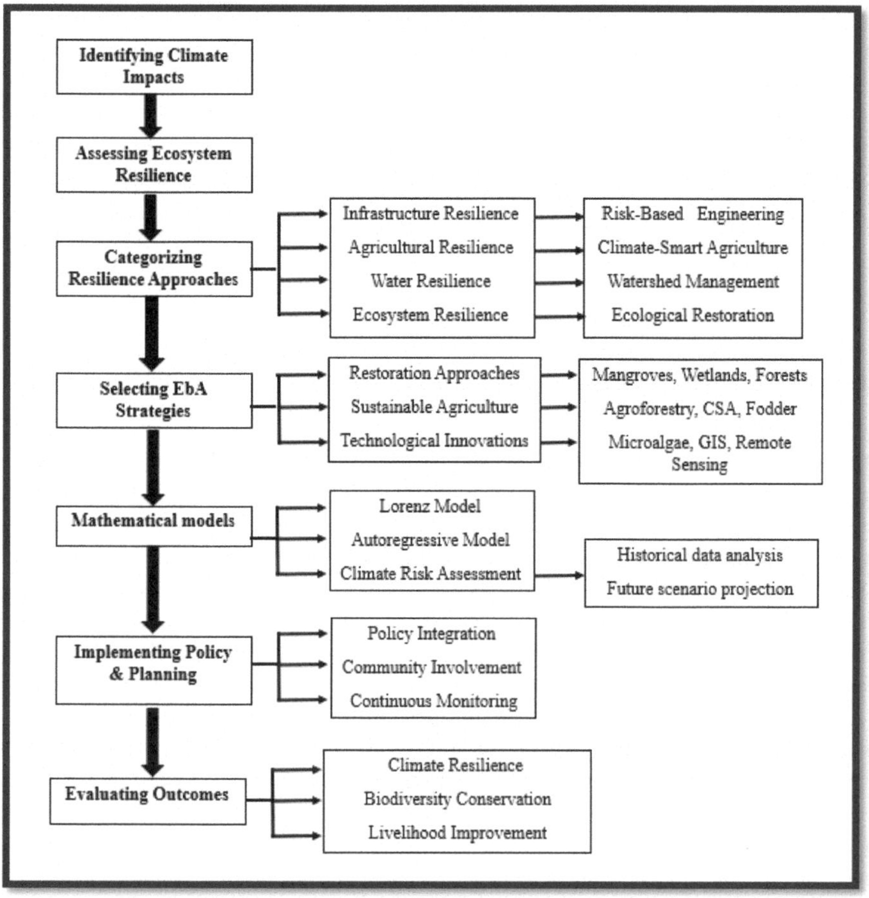

Fig. 14.2 Methodological framework

Measurement Approaches

Indicators of Ecosystem Health

Ecological metrics are vital tools for assessing and enhancing climatic resilience by quantifying the health and functionality of ecosystems. These metrics, such as biodiversity indices, habitat quality assessments, and ecosystem service valuations, provide critical insights into how ecosystems respond to climate stressors and their capacity to adapt and recover. For instance, metrics like species richness and evenness can indicate the stability and robustness of an ecosystem, while assessments of habitat connectivity and fragmentation help gauge its ability to support species migration and resilience (Harrison et al. 2023). Recent advancements have improved the precision of these metrics, incorporating high-resolution remote sensing and machine-learning techniques to monitor changes in real-time (Wang et al. 2024). This enhanced capability allows for more accurate predictions of ecosystem responses to climatic variations and facilitates targeted conservation efforts. By integrating these ecological metrics into climate resilience planning, we can better understand ecosystem vulnerabilities and design effective strategies to bolster their adaptive capacity (Smith and Jones 2023). Thus, ecological metrics inform resilience assessments and guide the implementation of adaptive management practices crucial for sustaining ecosystem functions amidst climate change.

Methods for Measuring Ecosystem Services

Measuring ecosystem services is crucial for understanding and enhancing climatic resilience, as it helps quantify ecosystems' benefits in mitigating and adapting to climate change. Recent methods for assessing these services include remote sensing, valuation techniques, and integrated modelling approaches. Remote sensing technologies, such as satellite imagery and aerial drones, enable high-resolution monitoring of land cover changes and ecosystem health, providing valuable data on services like carbon sequestration and water regulation (Miller et al. 2023). Valuation techniques, including contingent valuation and choice modelling, quantify the economic benefits of ecosystem services, which can inform cost–benefit analyses for climate resilience projects (Kumar and Kumar 2023). Additionally, integrated modelling approaches combine ecological data with socio-economic factors to predict how changes in ecosystem services impact climate resilience and human well-being (Chen et al. 2024). These methods collectively enhance our ability to monitor, evaluate, and manage ecosystem services, ensuring that resilience strategies are based on comprehensive and actionable data. Using these techniques, stakeholders can better align conservation and adaptation efforts with the ecosystem services critical for sustaining climatic resilience.

Tools and Technologies for Ecosystem Monitoring in Climatic Resilience

Tools and technologies for ecosystem monitoring are critical for assessing climatic resilience. Advanced remote sensing technologies, such as high-resolution satellite imagery and unmanned aerial vehicles (UAVs), provide detailed data on land cover and ecosystem changes (Miller et al. 2023). Additionally, in-situ sensors and Internet of Things (IoT) devices offer real-time data on environmental variables like temperature and soil moisture (Chen et al. 2024). Geographic Information Systems

(GIS) integrate these data to model and visualize ecosystem dynamics, enhancing our understanding of climate impacts and resilience strategies (Smith et al. 2023).

Assessment Techniques for Climatic Resilience

Ecosystem Vulnerability Assessments

Evaluating ecosystem vulnerability to climate impacts involves several vital methods. Vulnerability assessments typically use indicators such as species distribution models, habitat loss metrics, and climate exposure indices to gauge how ecosystems are affected by climate stressors (Harrison et al. 2023). Risk mapping tools, often integrated with Geographic Information Systems (GIS), visualize potential impacts and guide adaptation strategies (Wang et al. 2024). Additionally, scenario analysis helps predict future vulnerabilities under various climate scenarios (Smith and Jones 2023). These methods collectively provide a comprehensive understanding of ecosystem vulnerabilities, informing effective management and conservation efforts.

Integration of Social and Ecological Data

Integrating social and ecological data is crucial for assessing climatic resilience, as it provides a comprehensive understanding of human–environment interactions. Social data, such as community vulnerability and adaptive capacities, combined with ecological data, like biodiversity and ecosystem services, create a holistic framework for resilience assessment. This approach enables the identification of climate risks and the development of adaptive strategies that align with social needs and ecological stability. By bridging these data types, assessments can more effectively address the complexity of climate resilience (Biggs et al. 2015).

Prediction Models

Modeling Ecosystem Responses

Models for predicting ecosystem responses to climatic resilience typically include statistical, process-based, and hybrid models. Statistical models rely on historical data to identify trends and correlations in ecosystem responses to climate change. Process-based models simulate ecosystem processes, such as nutrient cycling and species interactions, to predict changes under various climate scenarios. Hybrid models combine statistical and process-based approaches to capture empirical patterns and underlying mechanisms. These models are crucial for anticipating ecosystem shifts and informing adaptive management strategies for climate resilience (Pereira et al. 2020).

Future Scenarios and Uncertainty

Future scenarios in climatic resilience predictions involve modeling various pathways based on socioeconomic, environmental, and technological changes. Scenarios outlined in the IPCC's Shared Socioeconomic Pathways (SSPs) explore different combinations of global policies, population growth, and economic development to predict how ecosystems and societies might respond to climate change. These scenarios enable researchers to estimate future risks and adaptation needs under diverse conditions (O'Neill et al. 2017).

However, uncertainty remains a significant challenge in predictions for climatic resilience. Sources of uncertainty include incomplete knowledge of climate system feedback, limitations in model resolution, and unpredictable human responses to climate policies. Additionally, ecological complexity—such as species interactions and nonlinear ecosystem responses—introduces further uncertainty in predicting how ecosystems will adapt to or resist climate stressors. These uncertainties can lead to various potential outcomes in resilience planning, from optimistic recovery scenarios to severe degradation under high-emission pathways. Addressing this uncertainty requires integrating multiple models, adopting adaptive management approaches, and continuously refining predictions as new data becomes available (Walker et al. 2020).

Future Perspectives

In recent decades, several emerging trends and innovative approaches have been poised to shape the field of ecosystem-based resilience. Advances in remote sensing, artificial intelligence, and big data analytics enhance the ability to monitor and model ecosystems with unprecedented precision. Technologies such as high-resolution satellite imagery and real-time sensor networks are revolutionizing the measurement of ecosystem health, while machine learning algorithms are improving predictive models of ecosystem responses to climate change. New research directions focus on integrating multi-scale and multi-disciplinary approaches to understand complex ecosystem dynamics better. This includes exploring the synergies between natural and human systems and incorporating social-ecological feedback mechanisms into resilience models. Collaborative research efforts that bridge ecological science, technology, and socio-economic analysis are expected to yield more profound insights into ecosystem resilience and adaptation strategies.

For policymakers and practitioners, there are key recommendations to enhance climate resilience through ecosystem-based approaches. First, adopting a holistic approach that integrates ecological, social, and economic factors into planning and decision-making processes is essential. Policymakers should promote cross-sectoral collaboration and invest in research and technology that support ecosystem monitoring and modeling. Practitioners are encouraged to use adaptive management strategies incorporating continuous feedback and learning. Strategies for integrating ecosystem-based approaches into climate resilience planning include developing robust frameworks that combine scientific research with local knowledge, fostering partnerships between governmental and non-governmental organizations, and promoting policies that incentivize sustainable practices. By embracing these innovative approaches and collaborative strategies, we can build more resilient and adaptive systems to address the challenges of climate change effectively.

Conclusion

Ecosystem-based approaches (EbA) are vital for building climate resilience, using biodiversity and ecosystem services to reduce the impacts of climate change and promote sustainable adaptation. These approaches, such as mangrove restoration and watershed management, offer dual benefits—protecting communities from climate impacts like storms and floods while promoting biodiversity conservation and improved livelihoods. Current EbA practices mainly focus on biodiversity and conservation, often overlooking the integration of these approaches into broader sectors such as livelihoods and disaster risk reduction, limiting their overall effectiveness and scope. To fully realize the potential of EbA, it is essential to mainstream these practices across various scales and contexts, incorporating governance and community participation to avoid top-down, ineffective planning. Technological advancements, such as remote sensing, advanced cultivation techniques, local engagement, and traditional knowledge, can further enhance EbA efforts. Additionally, adapting EbA strategies to local landscapes and integrating them into national policies and planning frameworks will be critical for long-term resilience. While challenges remain, particularly in data availability and governance, EbA offers a sustainable and adaptable pathway for climate resilience, benefiting human and natural systems in the face of global environmental changes.

References

Berkes F, Colding J, Folke C (2000) Rediscovery of traditional ecological knowledge as adaptive management. Ecol Appl 5:1251–1252. https://doi.org/10.2307/2641280

Biggs R, Schlüter M, Schoon ML (2015) Principles for building resilience: sustaining ecosystem services in social-ecological systems. Cambridge University Press

Chausson A, Turner B, Seddon D, Chabaneix N, Girardin CAJ, Kapos V, Key I, Roe D, Smith A, Woroniecki S, Seddon N (2020) Nature-based solutions can help cool the planet—if we act now. Nature 593:191–194. https://doi.org/10.1038/d41586-020-01241-2

Chen X, Zhang Y, Liu J (2024) Integrated modeling of ecosystem services for climate resilience planning. Environ Model Assess 29(2):321–335

Cohen-Shacham E, Walters G, Janzen C, Maginnis S (eds) (2016) Nature-based solutions to address global societal challenges. IUCN International Union for Conservation of Nature. https://doi.org/10.2305/IUCN.CH.2016.13.en

Folke C (2006) Resilience: the emergence of a perspective for social-ecological systems analyses. Glob Environ Chang 16(3):253–267. https://doi.org/10.1016/j.gloenvcha.2006.04.002

Gogoi N, Sarma A, Choudhury M, Samanta P, Faizan M, Sadhak S (2024a) Soil heavy metal pollution and ecological risk assessment in disturbed and undisturbed soil of Morigaon, Assam. Discover Environ 2(1):13. https://doi.org/10.1007/s44274-024-00039-z

Gogoi N, Choudhury M, Ul Hasan MS, Changmai B, Baruah D, Samanta P (2024) Soil organic carbon pool in diverse land utilization patterns in North-East India: an implication for carbon sequestration. Environ Dev Sustain, 1–27. https://doi.org/10.1007/s10668-024-05037-y

Harrison PA, Dunford R, Holman IP, Cojocaru G (2023) Assessing ecosystem resilience: metrics and methods. Ecol Ind 139:108255. https://doi.org/10.1016/j.ecolind.2022.108255

Holling CS (1973) Resilience and stability of ecological systems. Annu Rev Ecol Syst 4(1):1–23

IPCC (2014) Climate change 2014: impacts, adaptation, and vulnerability. In: Contribution of working group II to the fifth assessment report of the Intergovernmental Panel on Climate Change. Cambridge University Press

Kabisch N, Frantzeskaki N, Pauleit S, Naumann S, Davis M, Artmann M, Haase D, Knapp S, Korn H, Stadler J, Zaunberger K, Bonn A (2016) Nature-based solutions to climate change mitigation and adaptation in urban areas: perspectives on indicators, knowledge gaps, barriers, and opportunities for action. Ecol Soc 21(2):39. https://doi.org/10.5751/ES-08373-210239

Keesstra S, Nunes JP, Novara A, Finger D, Avelar D, Kalantari Z, Cerdà A (2018) The superior effect of nature-based solutions in land management for enhancing ecosystem services. Sci Total Environ 610–611:997–1009. https://doi.org/10.1016/j.scitotenv.2017.08.077

Kok MTJ, Meijer JR, van Zeist WJ, Verburg PH, van der Velde M (2021) Assessing ambitious nature conservation strategies within a 2-degree warmer and food-secure world. Nature Sustain 4(1):14–23. https://doi.org/10.1038/s41893-020-00555-3

Kremen C, Merenlender AM (2018) Landscapes that work for biodiversity and people. Science 362(6412):eaau6020. https://doi.org/10.1126/science.aau6020

Kumar D, Choudhury M (2021) Recognizing economic values of wetland ecosystem services: a study of emerging role of monetary evaluation of Chandubi ecosystem and biodiversity. In: Wetlands conservation: current challenges and future strategies, pp 87–110. https://doi.org/10.1002/9781119692621.ch5

Kumar P, Kumar M (2023) Economic valuation of ecosystem services: methods and applications. Ecol Econ 205:106232. https://doi.org/10.1016/j.ecolecon.2022.106232

Lavorel S, Locatelli B, Colloff MJ, Bruley E (2020) Co-producing ecosystem services for adapting to climate change. Glob Environ Chang 63:102120. https://doi.org/10.1016/j.gloenvcha.2020.102120

Macreadie PI, Costa MD, Atwood TB et al (2021) Blue carbon as a natural climate solution. Nat Rev Earth Environ 2(12):826–839. https://doi.org/10.1038/s43017-021-00262-6

Miller JR, Pettorelli N, Turner W et al (2023) Remote sensing for ecosystem service monitoring and climate resilience. Remote Sens Environ 272:112943. https://doi.org/10.1016/j.rse.2022.112943

Mobin S, Alam F (2017) Some promising microalgal species for commercial applications: a review. Energy Procedia 110:510–517. https://doi.org/10.1016/j.egypro.2017.03.140

O'Neill BC, Kriegler E, Riahi K, Ebi KL, Hallegatte S, Carter TR, van Vuuren DP (2017) The roads ahead: narratives for shared socioeconomic pathways describing world futures in the 21st century. Glob Environ Chang 42:169–180

Ostrom E (2009) A general framework for analyzing sustainability of social-ecological systems. Science 325(5939):419–422

Pelling M, O'Brien K, Matyas D (2015) Adaptation and transformation. Clim Change 133(1):113–127. https://doi.org/10.1007/s10584-014-1303-0

Pereira HM, Ferrier S, Walters M, Geller GN, Jongman RH (eds) (2020) Essential biodiversity variables: global observation needs for biodiversity change. Wiley. https://doi.org/10.1002/9781119386366

Reid WV, Mooney HA, Cropper A, Capistrano D, Carpenter SR, Chopra K (eds) (2005) Millennium ecosystem assessment: ecosystems and human well-being: synthesis. Island Press

Reid H, Phillips J, Heath M (2009) Natural resilience: healthy ecosystems as climate shock insurance. International Institute for Environment and Development (IIED)

Seddon N, Chausson A, Berry P, Girardin CAJ, Smith A, Turner B (2020) Understanding the value and limits of nature-based solutions to climate change and other global challenges. Philos Trans R Soc B Biol Sci 375(1794):20190120. https://doi.org/10.1098/rstb.2019.0120

Singh S, Singh P, Sharma A, Choudhury M (eds) (2022) Agriculture waste management and bioresource: the circular economy perspective. Wiley. https://onlinelibrary.wiley.com/doi/book/10.1002/9781119808428

Smith T, Jones A (2023) Incorporating ecological metrics into climate resilience planning. Climate Policy 23(5):750–764. https://doi.org/10.1080/14693062.2023.2161234

Smith T, Turner B, Patel S et al (2023) Advancements in GIS for ecosystem monitoring and climate resilience. J Spat Sci 68(1):15–28. https://doi.org/10.1080/14498596.2023.2184567

Spalding M, Ruffo S, Lacambra C, Meliane I, Hale L, Shepard C, Beck M (2014) The role of ecosystems in coastal protection: adapting to climate change and coastal hazards. Ocean Coast Manag 90:50–57. https://doi.org/10.1016/j.ocecoaman.2013.09.007

Walker B, Holling CS, Carpenter SR, Kinzig A (2020) Resilience, adaptability, and transformability in social–ecological systems. Ecol Soc 25(2):1–16. https://doi.org/10.5751/ES-08580-250201

Wang L, Zhang Q, Zhang Z (2024) Advancements in remote sensing for ecological monitoring and resilience assessment. J Environ Manage 331:117297. https://doi.org/10.1016/j.jenvman.2023.117297

Chapter 15
Sustainable Ecosystem Management for Strengthening Climate Resilience

Rwitabrata Mallick, Ayantika Banerjee, and Sandeep Poddar

Abstract In this research, the researchers have explored the fundamental nexus between ecosystem services and climate resilience, aiming to provide a comprehensive theoretical framework for understanding their interconnections. Ecosystem services, encompassing provisioning, regulating, cultural, and supporting services, are crucial in bolstering the resilience of communities and ecosystems to climate change impacts. Our literature synthesis identified key themes highlighting how various ecosystem services enhance adaptive capacity, reduce vulnerability, and improve overall resilience. However, gaps still need to be in our understanding, particularly in quantifying the economic value of ecosystem services and scaling up successful interventions across diverse socio-ecological contexts. Future research should address these gaps, leveraging interdisciplinary approaches to refine theoretical frameworks and inform policy and decision-making processes. Ultimately, integrating ecosystem services into climate resilience strategies offers a pathway toward building more adaptive, sustainable, resilient societies. By recognizing the inherent value of nature and harnessing its capacity to provide essential services, we can navigate the challenges posed by climate change more effectively and safeguard the well-being of current and future generations.

Keywords Ecosystem services · Sustainability · Climate resilience · Biodiversity

---

R. Mallick (✉)
Amity School of Life Science, Amity University Madhya Pradesh, Gwalior, India
e-mail: rmallick25088@gmail.com

A. Banerjee
Federal Institute for Material Research and Testing, BAM, Federal Ministry for Economic Affairs and Climate Protection, Berlin, Germany

S. Poddar
Lincoln University College, No. 12-18, Jalan SS6/12, Off Jalan Perbandaran, Petaling Jaya, Selangor, Malaysia

M. Choudhury et al. (eds.), *Sustainable Synergy: Harnessing Ecosystems for Climate Resilience*, Climate Change Management,
https://doi.org/10.1007/978-3-031-77957-2_15

Introduction

Conservation and restoration are two essential methods for preserving the natural and cultural legacy. Although they employ various strategies and techniques, they all aim to protect resources for the next generations. The preservation, sustainable use, and management of natural and cultural resources are the main goals of conservation. Preserving and enhancing these resources' authenticity and integrity is its core objective. Natural resource conservation, which entails the sustainable management of natural resources like forests, water, soil, and wildlife, is one category into which conservation approaches can be separated. The management of natural parks and protected areas, habitat preservation, and biodiversity protection are examples of conservation strategies. Intangible cultural assets like customs, dialects, and rituals are preserved along with intangible ones like historic structures, monuments, and artifacts. This is known as cultural heritage conservation. Some techniques include preventive conservation, cultural site documentation, and upkeep. Environmental conservation deals with more general ecological issues, including pollution, habitat destruction, and climate change. Community-based conservation initiatives, environmental policy development, and conservation planning are examples of strategies (Cassamo et al. 2023).

Restoring a degraded or damaged resource to its original state, or as near to it as feasible is the restoration process. It is frequently used when conservation measures alone cannot reverse damage or deterioration. Three restoration categories exist: environmental, cultural heritage, and ecological restoration. Restoring damaged, destroyed, or degraded ecosystems is known as ecological restoration. Techniques for restoration include replanting, restoring wetlands, and reintroducing natural species (Kumar and Choudhury 2021). Cultural heritage restoration aims to return historical structures, artwork, and artifacts to their original condition. Maintaining historical authenticity may entail cleaning, replacing any missing components, and applying conventional materials and methods. Using bioremediation, pollutant removal, and landscape rehabilitation, environmental restoration attempts to clean up contaminated ecosystems, including oil spills, restore contaminated soils, and revitalize urban areas.

To manage and protect resources to avoid harm, conservation is more comprehensive and preventive. Restoration addresses resources that have already been harmed and is more remedial. It has been noted that although restoration calls for specialized measures to undo the harm, conservation frequently entails continuing management techniques, monitoring, and policy execution. While conservation aims to prevent loss and restoration to reverse the damage, both endeavor to maintain resources for future generations. Ecological health must be balanced with economic and social considerations in conservation and restoration, considering the requirements and effects on human societies. Both professions face serious issues that require creative solutions and flexible management techniques. Prioritizing some projects and methods above others is necessary because of a need for more funding and human resources; this is frequently done depending on the degree of deterioration and the likelihood of successful outcomes (Giang and Khanal 2024).

Sustainable Land Use Practices

While addressing human requirements, sustainable land use practices are crucial to preserving ecosystem productivity and health. These methods seek long-term sustainability by balancing social, economic, and environmental goals. Agroforestry is the integration of forestry and agriculture techniques to provide more productive, diversified, and sustainable land-use systems. Incorporating trees and shrubs into livestock and agriculture systems has several advantages, including improved soil, biodiversity, regulated climate, and increased economic diversification (Hu et al. 2023).

Trees increase soil fertility and structure by fixing nitrogen and adding organic matter to the soil. They also improve the habitat for various species, promoting biodiversity. Trees produce timber, fruits, nuts, and other tree products, which generate additional revenue streams and help, reduce climate change by sequestering carbon (Gogoi et al. 2024a, b). The fundamental goals of conservation agriculture are to diversify plant species, minimize soil disturbance, and preserve a permanent soil cover.

The main features of land use management include

(a) Minimal tillage by reducing soil erosion and degradation.
(b) Crop rotation and diversity by improving soil health and reduces pest and disease cycles.
(c) Cover crops by protecting soil from erosion and improve soil fertility.

Sustainable grazing practices ensure that livestock production is compatible with the health of grasslands and other ecosystems. It is done by:

(a) Rotational grazing by moving livestock between pastures to prevent overgrazing.
(b) Proper stocking rates by maintaining an appropriate number of animals per unit area to balance forage availability and animal needs.
(c) Water management by ensuring sustainable water use for livestock and preventing water source contamination.

Biological, physical, cultural, and chemical treatments are combined in integrated pest management (IPM) to control pests in an economically and environmentally sustainable way. In this instance, biological management is achieved by using parasites or natural predators to manage pest populations. IPM reduces pest outbreaks using cultural techniques such as intercropping, crop rotation, and resistant types. In addition, there are barriers, traps, and manual pest elimination. Chemical control is achieved by prioritizing less hazardous solutions and using pesticides sparingly when necessary (Kong et al. 2023).

Organic farming forgoes synthetic pesticides and emphasizes natural processes to preserve soil fertility and ecological balance (Gogoi et al. 2024a, b). These methods include enriching soil with compost, manure, and green manures, using organic pesticides and natural predators, building resilience, and lowering reliance on single crops. The goal of permaculture, a comprehensive approach to land use planning, is to establish self-sufficient and sustainable agricultural ecosystems. Designing for sustainability entails planning land use to resemble natural ecosystems, maximizing outputs while minimizing energy inputs, and recycling waste materials into system resources.

Restoring degraded areas, improving carbon sequestration, and promoting biodiversity depend on afforestation, planting new trees, reforestation, or planting native forests. Involving local populations in the development and upkeep of forest initiatives is just as vital as planting native trees compatible with the local environment. Harvesting forest products in a way that does not jeopardize the forest's health is essential. Urban planning that includes green spaces improves both the environment and human well-being. This can be accomplished by establishing green spaces in urban areas to enhance air quality and offer leisure places, adding vegetation to buildings to lessen urban heat islands and improve insulation, and promoting urban agriculture and community involvement (Ma et al. 2022).

Watershed management pertains to the sustainable administration of water resources throughout a watershed, tackling land use patterns that impact the availability and quality of water. This can be accomplished by reestablishing wetlands to enhance habitat and water filtration, planting vegetation beside water bodies to stabilize banks and filter runoff, and using effective irrigation methods to save water. Promoting sustainable behaviors requires effective land use planning and policies. To balance needs and safeguard resources, land use planning entails identifying specific areas for development, conservation, and agriculture and offering financial and technical support for sustainable practices (Choudhury et al. 2021). It is evident that including local communities in the process of making decisions can contribute to the improvement of their needs and values.

Climate-Smart Agriculture

Climate-Smart Agriculture (CSA) is an integrated approach to managing landscapes—cropland, livestock, forests, and fisheries—that aims to achieve three main objectives: (a) sustainably increasing agricultural productivity and incomes, (b) adapting, and building resilience to climate change, and (c) reducing or removing greenhouse gas emissions.

Smart agriculture includes creating and using crop types that are resistant to harsh weather, as well as cultivating crops with improved nutritional profiles to combat malnutrition. Growing cover crops to prevent soil erosion, boost fertility, and control weeds is more critical than minimally disturbing the soil to enhance its structure and increase organic matter. To improve soil fertility and moisture retention, organic matter must be added. Effective irrigation methods that conserve water, lessen runoff, and collect and store rainwater for use in agriculture during dry spells are also necessary for climate-wise agriculture (Ma et al. 2021).

The need for low-carbon transformation, urban ecological management, and sustainable human development is growing because of the numerous hazards and problems that global climate change, rapid urbanization, and abrupt economic development present to humanity. Currently, there are many different urban management models, but most of them concentrate on one-dimensional, problem-oriented measures.

Studies found that urban ecological management's development includes four stages

(a) the germination stage,
(b) The transitional stage,
(c) The formative stage, and
(d) The prosperity stages.

It has also been observed that, urban ecological management models should incorporate following perspectives of urban planning

- Industrial transformation,
- Organizational model,
- Environmental protection
- Institution system.

The framework can offer a solid scientific foundation for improving urban ecological management. All the analyses and findings have the potential to further practical low-carbon transformation initiatives, regional sustainable development initiatives, and scholarly studies. To maximize irrigation schedules and cultivate crops that need less water or are resistant to water stress, which can play a key role in the future, new instruments that monitor soil moisture levels must be developed.

To lower risks and improve biodiversity and ecosystem services, smart agriculture advocates integrating trees and shrubs into agricultural landscapes and cultivating many crops in the same area concurrently or in succession. It has been noted that integrating livestock and crop farming can improve resilience and optimize resource utilization. However, adding fish aquaculture to current agricultural systems can increase food security and diversify sources of income. Farmers may more effectively plan their activities with access to reliable weather forecasts. Farmers can optimize farming operations and receive financial insurance against crop loss due to climate-related events by receiving customized advice based on climate predictions (Madhubhashini 2024).

Breeding animals with increased resistance to heat and illness is one way to achieve sustainable livestock management or SLM. Efficient feed must also be used with high-quality feed and feeding strategy optimization to increase productivity and lower methane emissions. Moving cattle across pastures has been found to reduce overgrazing and encourage regrowth by incorporating trees into pastureland to offer shade and enhance soil health. To lessen dependency on fossil fuels, the current study has also discovered that growing crops expressly to produce biofuel, utilizing solar energy to power irrigation systems, utilizing wind energy for various farm operations, and converting agricultural waste into biogas for cooking and heating can all be implemented. Sustainable livestock management will strengthen agricultural extension services to disseminate CSA knowledge and practices by involving farmers, communities, governments, and NGOs in implementing CSA strategies. This will be achieved by investing in agricultural research to develop innovative CSA technologies and offering financial support for adopting CSA practices. This will guarantee long-term agricultural productivity and food security and improve agricultural systems' capacity to resist and recover from climate shocks. By using diverse and resilient farming systems, the method lowers greenhouse gas emissions, increases carbon sequestration, and improves farm incomes and livelihoods (Nwokolo et al. 2023).

Protecting Coastal Ecosystems

To preserve biodiversity, sustain livelihoods, and lessen the effects of climate change, coastal ecosystems must be protected. Numerous ecological, economic, and social advantages are offered by coastal habitats, which include seagrass beds, salt marshes, coral reefs, and mangroves. By putting in place legislative frameworks and enforcement measures to stop deforestation and degradation, existing mangroves can be protected. It is possible to improve the restoration of these vital habitats by planting mangroves in degraded or cleared regions. Involving local communities in conservation initiatives through education and reforestation operations is crucial. Marine protected areas (MPAs) can extensively safeguard coral reefs from harmful activities, pollution, and overfishing. Accelerating methods like transplantation and coral gardening are necessary to repair damaged reefs. The rate of pollution can be decreased by lowering runoff from urban and agricultural regions, which contains contaminants detrimental to coral reefs. A comprehensive strategy that balances social, economic, and environmental objectives in coastal areas is called integrated coastal zone management, or ICZM.

Putting land use planning and zoning laws are into practice to stop overdevelopment and habitat degradation and create vegetated buffer zones to shield coastal regions from pollution and erosion (Pörtner et al. 2023).

Ecosystem management includes

- Educating visitors about the value of coastal ecosystems,
- Promoting actions that preserve these settings, and
- Promoting responsible tourism that reduces its adverse effects on the environment and supports conservation efforts.

In addition to lowering marine pollution, it will also manage trash by implementing efficient mechanisms for recycling, collecting, and disposing of waste to prevent litter and plastic contamination in coastal areas. Reducing nutrient and chemical pollution can be achieved by managing runoff from both urban and agricultural areas using green infrastructure, such as permeable pavements and rain gardens.

It is creating and upholding plans for responding to oil spills to reduce their adverse effects on coastal ecosystems, coordinating undersea and coastal clean-ups to eliminate trash and lessen its effect on marine life, and encouraging the decrease of single-use plastics using alternatives, prohibitions, and public awareness initiatives. Maintaining and restoring salt marshes, sea grass beds, and mangroves—all of which absorb significant amounts of carbon dioxide and contribute to reducing global warming and creating carbon offset schemes that support preserving and rehabilitating blue carbon ecosystems. Erecting seawalls and raised structures are examples of infrastructure that can survive adverse weather conditions and sea level rise—putting in place natural barriers to prevent erosion and storm surges, such as oyster reefs and living shorelines (Raj et al. 2024).

Motivating the local population to participate actively in preserving and managing coastal environments is the need of the hour. It is also significant to incorporate local and indigenous knowledge into conservation initiatives and create initiatives to teach the public—youth in particular—about the value of coastal habitats and the need to preserve them. The process also needs collaboration to advance sustainable practices and conservation initiatives with various stakeholders, such as governments, NGOs, companies, and local people. Monitoring the condition of coastal habitats and identifying hazards through routine evaluations is essential. It investigates and creates practical methods for restoring deteriorated coastal environments and monitoring shifts in the diversity and number of organisms to find early indicators of environmental stress. Evaluating water quality regularly to find and fix contamination sources is required.

It is essential to uphold and enhance legislation to preserve coastal habitats from pollution and overuse. The process of participating in international conventions and agreements, including the Convention on Biological Diversity (CBD), is devoted to conserving the sea and the shore. Putting into effect and upholding sustainable fishing methods to stop overfishing and save marine biodiversity is significant. Distributing space for various uses in a way that reduces are disputes and adverse effects on the environment through the application of maritime spatial planning needs to be understood. It can be assured that coastal ecosystems will continue to offer their essential services to future generations by implementing various ideas and practices (Salliou and Stritih 2023).

Water Resource Management

Effective management of water resources is crucial to guarantee the sustainable use of water for diverse human requirements while preserving the well-being of aquatic ecosystems. Integrated Water Resources Management (IWRM) assists in the holistic management of water resources by considering the demands of multiple stakeholders and all facets of the water cycle. Communities, businesses, governments, and other stakeholders participate in management's decision-making process. The ability of future generations to meet their water needs must not be jeopardized by excessive water use. Water-saving technologies include installing dual-flush toilets, low-flow fixtures, and energy-efficient appliances. It is also critical to routinely inspect and rectify plumbing system leaks to stop water loss (Sherpa 2024).

Industrial operations should have procedures for recycling and reusing water, and frequent water audits should be carried out to find areas where water consumption efficiency could be increased. In addition to routinely measuring groundwater levels and quality to identify trends and problems early on, other technologies and practices that reduce water usage and pollution should be adopted. Rainwater can be captured and stored to replenish groundwater supplies using implementable techniques such as injection wells and recharge basins and mapping aquifers to determine their area, capacity, and recharge rates.

Ecosystem management can be achieved by implementing sedimentation control and reservoir capacity maintenance measures and preventing development in regions vulnerable to flooding using land-use planning. In the process of reducing the risk of flooding, it is essential to build retention basins, floodwalls, and levees and put emergency response plans, flood forecasts, and early warning systems into action. Controlling and keeping an eye on discharges from municipal, agricultural, and industrial sources is essential to lessen pollution. Using best management practices (BMPs) will help manage runoff from urban areas, agriculture, and other diffuse sources and treat wastewater to the highest standards, using technologies like membrane filtration, bioreactors, and nutrient removal. It is needed to encourage the repurposing of cleaned wastewater for groundwater recharge, industrial processes, and irrigation (Stritih and Salliou 2024). The public must be made aware of water-saving techniques, the value of water conservation, and the use of drip or sprinkler irrigation to maximise efficiency and minimise water use to achieve water resource management (Fig. 15.1). Strategies like mulching and soil moisture sensors can help to maximise irrigation schedules and support the growth of crops that use less water and are more drought tolerant. Figure 15.1 shows water resource management techniques.

Establish and implement water quality standards to safeguard ecosystems and public health. Check water bodies regularly to monitor water quality and pinpoint sources of pollution. Preserve and restore wetlands to improve water filtration, flood control, and biodiversity. Restore rivers' ecosystems and natural processes to enhance ecological health and water quality. When allocating water, consider the demands of ecosystems to sustain their health and function. Mimic the flow patterns found in nature to aid aquatic organisms' life cycles. Water infrastructure should be designed and updated to endure harsh weather and shifting climatic conditions. Create management strategies that are adaptive to changing and unpredictable climate conditions. Use climate models and scenarios to plan for future water availability and demand. Create plans for managing droughts and implement policies to improve water availability during dry spells. Involve local businesses, communities, and other interested parties in the planning and decision-making process for water management. Give stakeholders the tools and training they need to become more capable of contributing to water management. Run initiatives to increase public knowledge of sustainable

Fig. 15.1 Water resource management

behaviors, pollution avoidance, and water conservation. Include water education in the curriculum to help pupils understand the value of water resources (Trott et al. 2024).

Integrated Landscape Management

Managing land, water, and other natural resources in a coordinated manner to accomplish various goals, including livelihoods, biodiversity preservation, and sustainable agriculture, is known as integrated landscape management (ILM) (Fig. 15.2). This holistic approach balances environmental, social, and economic goals across entire landscapes by involving local groups in decision-making processes, including farmers, landowners, and indigenous peoples. Through this approach, collaborations between various industries, including forestry, conservation, water management, and agriculture, can be fostered, and a shared vision for the landscape that considers stakeholders' multiple objectives and interests can be developed. Additionally, it will support the development of coordinated plans that address resource management, conservation, land use, and development objectives. Incorporating trees into agricultural landscapes can yield several benefits, including increased biodiversity, low tillage, crop rotation, and cover crops to support soil health and production. The ultimate goal of the entire process should be the long-term productivity and health of forest ecosystems. The process should create and administer protected areas to conserve important species and habitats by connecting disparate habitats and facilitating the movement of wildlife and genetic diversity (Viñals et al. 2023).

Trees should be planted in deforested or degraded areas to promote carbon sequestration and restore ecosystems. Additionally, restoring wetlands will improve flood control, biodiversity, and water quality. Watershed-based resource management can address problems with water distribution, amount, and quality. Utilizing technology like drip irrigation and soil moisture sensors to optimize water consumption in agriculture will enable the establishment of vegetated buffers along waterways to filter

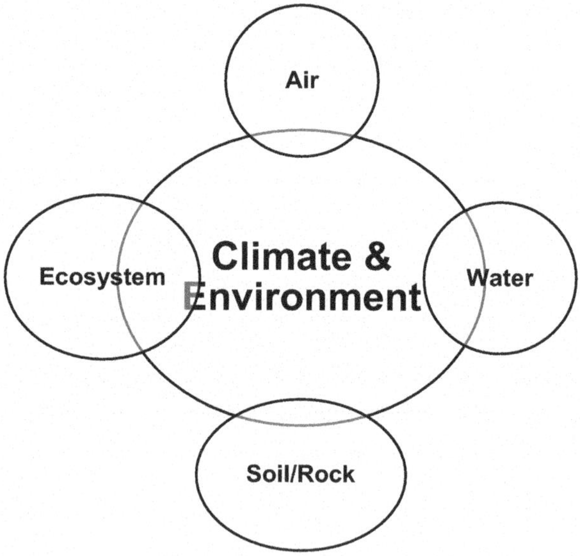

Fig. 15.2 Landscape management

pollutants and stabilize banks. Rainwater harvesting devices can collect and store water during dry spells. Figure 15.2 shows components landscape management.

Natural disaster risk and damage can be decreased by implementing agricultural methods that are more resilient to climate variability and change. Examples of these practices include floods and droughts. Reduced tillage and cover crops are two techniques that should improve soil carbon storage. The development of diverse income sources, such as agroforestry, ecotourism, and non-timber forest products, as well as strengthening value chains for sustainable products, will be encouraged by projects that protect forests and increase carbon sequestration to mitigate climate change. These actions will also increase market access and benefit local communities economically. Proper training should be provided on sustainable practices, business skills, and market opportunities by strengthening agricultural extension services to support farmers in adopting sustainable practices.

Community Engagement and Capacity Building

Projects for sustainable development, conservation, and managing natural resources must include community involvement and capacity building (Fig. 15.3). These strategies guarantee that local populations have the abilities and information required to manage resources sustainably and participate actively in decision-making processes. Inclusive planning incorporates community people in the planning and decision-making processes to ensure that their needs and viewpoints are considered. Focus groups, open forums, and workshops can help accomplish this. Community members are surveyed regarding their resources, needs, and priorities using the Participatory Rural Appraisal (PRA) technique. This calls for techniques like problem rating, seasonal calendars, and mapping. Cultural sensitivity plays a crucial role in respecting local customs, traditions, and knowledge by engaging with community leaders and elders to build trust and foster collaboration. There should be open and honest communication about project goals, processes, and outcomes by providing regular updates and opportunities for feedback. Figure 15.3 shows public participation and community engagement.

It is essential to find and assist local leaders who can promote community engagement through community-based natural resource management (CBNRM), where local people are actively involved in resource management and decision-making. Developing community members' organizational and leadership capacities is crucial for improving the prevailing situation. Communities should communicate with one another to exchange best practices and experiences. To encourage community members to invest in sustainable businesses, community involvement should offer and support resilient and alternative livelihoods, including agroforestry, ecotourism, and artisanal crafts. Supporting Community-Based Organisations (CBOs) is essential because they may oversee resources, speak for the community's interests, and encourage the formation of associations and cooperatives to strengthen group dynamics and bargaining power.

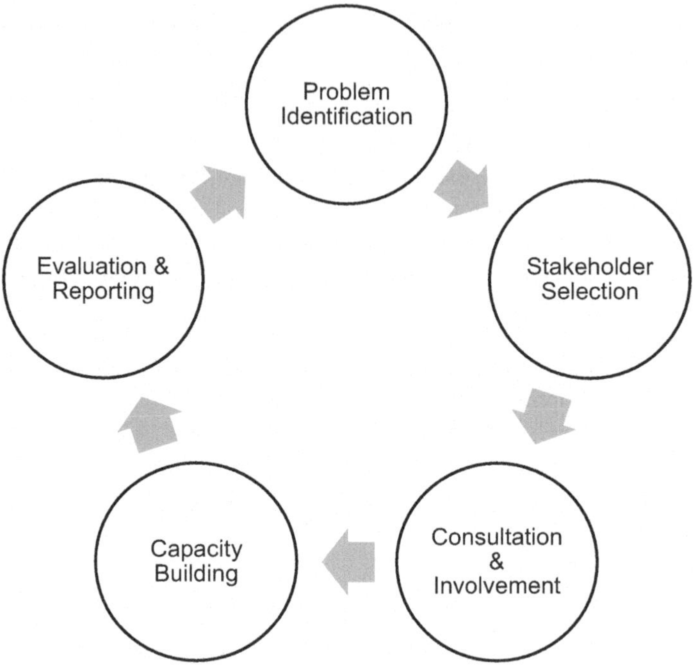

Fig. 15.3 Public participation and community engagement

Ecosystem management should always include educational campaigns to increase public awareness of the value of conservation and sustainable resource management (WHO 2023). These campaigns should use a variety of communication channels, including local newspapers, social media, and community radio, to engage the community and spread knowledge. In this regard, a Kurseong Tea Gardens case study can be considered. The community-based Society in Kurseong is an autonomous and independent body. Its primary function is to preserve and protect the natural, historical, and cultural heritage of Kurseong. It should be formed and organized diligently so that it may act as a preferred destination for visitors, and it provides opportunities to promote Community Ecotourism (CBET) in the Kurseong hill area, mainly using the latest approaches of Participatory Forestry Management for sustainable development of the forests, thereby generating revenue for the hill area. Regardless of the geographical location, harmony is to be fostered between the people, environment, conservation, and development. This Society would partner with the civil Society (local communities, NGOs, academic institutions, and private enterprises/businesses) and the State (Forest Department, the Tourism Department, and other related departments). The involvement of local communities would support their livelihood and consequently create a direct stake in the conservation of local culture, ecology, and environment. It also means less dependence upon natural resources, as it amounts to the removal of produce from these areas.

Moreover, there would be more significant incentives for conserving local traditions, cultural heritage, and environment, as these would be seen as an aid to the local economy rather than being independent of it. To encourage ongoing participation and effort, people should acknowledge and celebrate the accomplishments of community members and groups. They should also share success stories and case studies with other communities and stakeholders to educate and inspire them. Ultimately, this will support community-led watershed management initiatives that have increased agricultural output, climate change resistance, and water availability.

It is recommended that major water-saving strategies be implemented to improve water resource management sustainably. Effective water resource management is crucial for securing long-term water supplies, preserving ecosystems, and promoting economic development. Integrated Water Resources Management (IWRM) can be a comprehensive solution for coordinating the development and management of water, land, and related resources to maximize economic and social welfare while preserving ecological sustainability. Water conservation and efficiency should be promoted in agriculture, industry, and the home by implementing technologies and practices that minimize waste. Sustainable agricultural techniques and efficient irrigation systems such as drip or sprinkler irrigation should be encouraged to reduce water use in agriculture and promote the development of drought-resistant and water-efficient crops. Governments and organizations may ensure the long-term management of water resources by implementing these guidelines, which will benefit both human and environmental health.

Policy and Governance

Successful governance and policy frameworks are essential for managing ecosystems. They offer the framework and rules required to balance the development requirements and preserve natural resources. To ensure cogent and coordinated management, it is critical to design policies that combine many sectors, including forestry, agriculture, water, and biodiversity. To guarantee comprehensive and sustainable management practices, national policies must appropriately align with global frameworks such as the Sustainable Development Goals (SDGs). Laws like the Wildlife Protection Act and the Water Act that prevent ecosystem damage must be passed and upheld. Protected areas and biological corridors are two conservation requirements that must be balanced with development by applying zoning and land use rules.

Strengthening Governance Structures is a prerequisite for discussing ecosystem management. Establishing or bolstering organizations like forestry departments and environmental protection agencies devoted to ecosystem management is one way to do this. To guarantee integrated administration and prevent competing policies, it is also necessary to encourage coordination across the many government departments and agencies. Local empowerment can play a significant role by giving local governments and community management authority and the resources they need to manage ecosystems sustainably. Initiatives for community-based natural resource management (CBNRM), which provide local people the authority to manage and profit from their resources, are always necessary (Yin et al. 2023a, b).

Platforms for discussion and decision-making involving the public sector, NGOs, the commercial sector, and community members must be established. Encouraging public involvement in policy creation and implementation processes via public hearings, consultations, and participatory planning is crucial. Indigenous and local knowledge can be integrated to increase the efficacy and cultural relevance of ecosystem management techniques and policies. In managing ecosystems, it is critical to acknowledge and uphold the rights of indigenous peoples and local communities, particularly those related to land tenure and resource access. Establishing robust monitoring and enforcement procedures is necessary to guarantee adherence to environmental laws and regulations.

The current study ensures that a mix of fines for non-compliance and rewards for voluntary conservation efforts is used to encourage policy adherence. It will assist in educating and training public servants, community leaders, and other interested parties on legislative obligations and best practices for managing ecosystems. Resources and technical support are also required to execute ecosystem management strategies at all levels. Implementing Payments for Ecosystem Services (PES) schemes is necessary, in which individuals who benefit from ecosystem services—such as water users or operators of tourism businesses—account for the upkeep of those services. It can be done by funding conservation and sustainable management initiatives via green bonds and environmental funds. It is critical to enter into international partnerships and agreements to leverage money from international donors, development banks, and charitable organizations for ecosystem conservation projects and to get funding and technical support for ecosystem management activities (Yin et al 2023a, b).

Sustainable ecosystem management can be aided by creating and applying indicators to track ecosystem functionality and health, such as soil health, water quality, and biodiversity. Monitoring the socioeconomic effects of ecosystem management—such as enhanced livelihoods and community well-being—as well as putting in place systems for ongoing feedback and learning how to modify management strategies and policies in response to monitoring findings are crucial. It is obvious that a culture of constant development should be promoted to adapt to new information and problems.

Policies that increase ecosystems' and communities' resistance to the effects of climate change can be developed and implemented through ecosystem management. Additionally, it can support methods like reforestation, sustainable agriculture, and renewable energy that lower greenhouse gas emissions. Identifying and mitigating ecological and community vulnerabilities to natural disasters can be significantly aided by conducting risk assessments. Resilient infrastructure and ecosystem-based strategies require investment to lower the risk and damage caused by natural catastrophes. It is necessary to ensure that information about ecosystem health and management techniques is readily available to the public. Ecosystem management includes:

- Keeping lines of communication open with all parties involved.
- Giving frequent updates.
- Offering opportunities for feedback.

It is also essential to set up channels for stakeholders to voice complaints and concerns, ensure that these are swiftly and equitably resolved, and regularly carry out independent audits and assessments of ecosystem management programs and policies (Zhang et al. 2024).

Conclusion

Sustainable management of natural and cultural resources must include conservation and restoration. Both restoration and preservation are necessary to ensure the longevity and health of the planet's unique legacy as restoration attempts to heal and recover. In contrast, conservation seeks to protect and sustain. Policymakers and governance institutions can effectively manage ecosystems by implementing these policies and practices, striking a balance between environmental conservation, sustainable development, and community well-being. It is possible to accomplish sustainable development objectives, strengthen ecosystem resilience, and improve community livelihoods by implementing integrated landscape management strategies.

A holistic approach guarantees that resource and land use is managed to meet social, economic, and environmental goals throughout landscapes. Social well-being, financial viability, and long-term ecological balance can all be attained by applying sustainable land use methods. A comprehensive and integrated strategy that balances human and environmental needs is necessary to manage water resources effectively. Implementing various policies and procedures can guarantee the sustainable use and preservation of priceless water resources. Projects can develop strong, resilient communities capable of managing their resources sustainably and enhancing their living standards by implementing various conservation measures.

Community involvement and capacity building are crucial for any development or conservation project to be sustainable over the long term. Climate-smart agriculture must be implemented to meet the problems presented by climate change and ensure sustainable agricultural development. Drones, sensors, and data analytics are a few examples of technology that should be used to optimize agricultural methods and minimize resource consumption. Renewable energy sources like solar and wind power should be incorporated into landscape management to lessen dependency on fossil fuels. Utilise sensors and Internet of Things (IoT) devices to monitor water usage, find leaks, and improve irrigation. Utilise data analytics to control supply, forecast water demand, and enhance decision-making. Water pumps, treatment facilities, and other infrastructure can be powered by solar energy. Install hydropower systems on a small scale to produce energy from water flows. Farmers can increase their production, resilience, and environmental sustainability by using CSA methods.

Ecosystem management is crucial for reducing the effects of a climate change scenario and promoting sustainable socioeconomic growth.

References

Cassamo CT, Draper D, Romeiras MM, Marques I, Chiulele R, Rodrigues M, Stalmans M, Partelli FL, Ribeiro-Barros A, Ramalho JC (2023) Impact of climate changes in the suitable areas for *Coffea arabica* L. production in Mozambique: agroforestry as an alternative management system to strengthen crop sustainability. Agri Ecosyst Environ 346:108341

Choudhury M, Sharma A, Singh P, Kumar D (2021) Impact of climate change on wetlands, concerning Son Beel, the largest wetland of North East, India. In: Global climate change. Elsevier, pp 393–414. https://doi.org/10.1016/B978-0-12-822928-6.00006-X

Giang PQ, Khanal R (2024) What next for marine ecosystem management in Vietnam: assessment of coastal economy, climate change, and policy implication. Environ Res Commun 6(2):025002

Gogoi N, Choudhury M, Ul Hasan MS, Changmai B, Baruah D, Samanta P (2024). Soil organic carbon pool in diverse land utilization patterns in North-East India: an implication for carbon sequestration. Environ Dev Sustain, 1–27. https://doi.org/10.1007/s10668-024-05037-y

Gogoi N, Sarma A, Choudhury M, Samanta P, Faizan M, Sadhak S (2024b) Soil heavy metal pollution and ecological risk assessment in disturbed and undisturbed soil of Morigaon, Assam. Discover Environ 2(1):13. https://doi.org/10.1007/s44274-024-00039-z

Hu Y, Gong J, Li X, Song L, Zhang Z, Zhang S, Zhang W, Dong J, Dong X (2023) Ecological security assessment and ecological management zoning based on ecosystem services in the West Liao River Basin. EcolEng 192:106973

Kong Z, Ling H, Deng M, Han F, Yan J, Deng X, Wang Z, Ma Y, Wang W (2023). Past and projected future patterns of fractional vegetation coverage in China. Sci Total Environ 902:166133

Kumar D, Choudhury M (2021) Recognizing economic values of wetland ecosystem services: a study of emerging role of monetary evaluation of Chandubi ecosystem and biodiversity. In: Wetlands conservation: current challenges and future strategies, pp 87–110. https://doi.org/10.1002/9781119692621.ch5

Ma S, Li Y, Zhang Y, Wang LJ, Jiang J, Zhang J (2022) Distinguishing the relative contributions of climate and land use/cover changes to ecosystem services from a geospatial perspective. Ecol Ind 136:108645

Ma S, Wang LJ, Jiang J, Chu L, Zhang JC (2021) Threshold effect of ecosystem services in response to climate change and vegetation coverage change in the Qinghai-Tibet Plateau ecological shelter. J Clean Prod 318:128592

Madhubhashini ETS (2024) Smart solutions for mitigating climate change impacts on coastal waters and resources in Sri Lanka. YSF Thematic Public 2024:191

Nwokolo SC, Meyer EL, Ahia CC (2023) Credible pathways to catching up with climate goals in Nigeria. Climate 11(9):196

Pörtner HO, Scholes RJ, Arneth A, Barnes DKA, Burrows MT, Diamond SE, Duarte CM, Kiessling W, Leadley P, Managi S, McElwee P, Midgley G, Ngo HT, Obura D, Pascual U, Sankaran M, Shin YJ, Val AL (2023) Overcoming the coupled climate and biodiversity crises and their societal impacts. Science 380(6642):eabl4881

Raj A, Jhariya MK, Banerjee A, Meena RS, Kumar S, Devi A (2024) Agroforestry and ecosystem services. In: Agroforestry for carbon and ecosystem management. Academic Press, pp 205–221

Salliou N, Stritih A (2023) In complexity we trust learning from the socialist calculation debate for ecosystem management. Environ Res Lett 18(5):051001

Sherpa TO (2024) Integration of urban ecosystem-based adaptation in Nepal: a policy landscape analysis. PLoS ONE 19(1):e0297786

Stritih A, Salliou N (2024) Reply to comment on 'In complexity we trust: learning from the socialist calculation debate for ecosystem management.' Environ Res Lett 19(1):018001

Trott CD, Weinberg AE, Frame SM, Jean-Pierre P, Even TL (2024) Civic science education for youth-driven water security: a behavioral development approach to strengthening climate resilience. Int J Behav Dev 48(2):145–155

Viñals E, Maneja R, Rufí-Salís M, Martí M, Puy N (2023) Reviewing social-ecological resilience for agroforestry systems under climate change conditions. Sci Total Environ 869:161763

Wang Y, Liu Y, Shan L, Du J, Liu Y, Li T, Cui X (2024) Ecosystem management and sustainable livelihoods in drylands. In: Dryland social-ecological systems in changing environments. Springer Nature, Singapore, pp 139–157

World Health Organization (2023) Addressing climate change: supplement to the WHO water, sanitation, and hygiene strategy 2018–2025. World Health Organization

Yin C, Pereira P, Zhao W, Barcelo D (2023) Natural climate solutions. The way forward. Geogr Sustain

Yin J, Zhang C, Xue Y, Ji Y, Xu B, Ren Y (2023b) Potential impacts of ocean warming on the trophic control of a threatened marine ecosystem. Prog Oceanogr 211:102976

Zhang P, Jia M, Wei X (2024) Sustainable water–ecosystem management using the Bayesian network and variable relationship analysis. Water Supply, ws2024109

Chapter 16
Greening the Future: Restoring and Conserving Biodiversity for Generations to Come

Ratan Chowdhury, Ambarish Bhuyan, Abhilash Bhattacharjee, and Ravi Rajak

Abstract Restoration and biodiversity conservation are crucial for the planet's long-term sustainability, requiring diverse strategies and collaborative efforts. Restoration involves habitat rehabilitation and reintroduction to create functional ecosystems that support rich biodiversity. Innovative methods such as genetic rescue and assisted migration enhance populations and the genetic diversity of threatened species. Conservation extends beyond protected areas, adopting landscape-level planning and ecosystem-based management to balance ecological and socio-economic needs. Community-based conservation empowers local communities to manage their natural resources, fostering ownership and responsibility. Recognizing the intrinsic value of nature and the interconnectedness of all life is central to effective biodiversity conservation. Conservation ethics and environmental education are vital in fostering a conservation ethos and deeper appreciation for the natural world, encouraging individuals to advocate for biodiversity at various scales. However, habitat loss, pollution, invasive species, climate change, and unsustainable resource use complicate conservation efforts. Addressing these challenges requires a concerted effort from governments, NGOs, academia, industry, and civil society. Collaboration and partnerships are essential for resource mobilization, knowledge sharing, and effective implementation of conservation measures. By embracing innovative approaches, fostering collaboration, and nurturing a reverence for nature, we can work towards a future where biodiversity thrives and humanity coexists harmoniously with the natural world.

R. Chowdhury (✉)
Department of Botany, Rangapara College, Rangapara, Assam 784505, India
e-mail: ratan@rcr.ac.in

A. Bhuyan · A. Bhattacharjee
Department of Botany, D.H.S.K. College, Dibrugarh, Assam 786001, India

R. Rajak
Department of Economics, The University of Bristol, Bristol, UK

M. Choudhury et al. (eds.), *Sustainable Synergy: Harnessing Ecosystems for Climate Resilience*, Climate Change Management,
https://doi.org/10.1007/978-3-031-77957-2_16

Keywords Ecological restoration · Collaborative conservation · Conservation ethics

Introduction

The current era marks a crucial turning point in the history of life on Earth—a biological renaissance where the fate of countless species is hangs in the balance. Biodiversity, the variety of life forms on Earth, is a marvel of nature and the foundation of human existence. Biodiversity encompasses the rich tapestry of life on our planet, encompassing various levels of organization, from genes to ecosystems. It includes species, genetic, and ecosystem diversity, each playing a vital role in upholding ecological balance and resilience. Biodiversity is not just about the number of species but their interactions, adaptations, and evolutionary potential. The significance of biodiversity cannot be emphasized enough. It underpins ecosystem services essential for human well-being, such as clean air and water, pollination, climate regulation, and soil fertility. Furthermore, biodiversity provides aesthetic, cultural, and recreational values, enriching our lives beyond mere economic utility.

Current State of Biodiversity Loss

Despite its significance, biodiversity is encountering unparalleled threats primarily driven by human activities. Habitat destruction, pollution, climate change, overexploitation of resources, and invasive species has accelerated species extinctions to alarming rates. The present rate of species extinction is estimated to be 1000 to 10,000 times greater than the natural background extinction rate (IPBES, 2019a, b). The decline of biodiversity diminishes the intricate web of life and weakens ecosystem stability and resilience. It disrupts crucial ecological processes, resulting in cascading effects on human societies, such as reduced agricultural productivity, increased vulnerability to natural disasters, and compromised access to essential resources (Gogoi et al. 2024a, b).

Need for Restoration and Conservation Efforts

In the face of escalating biodiversity loss, urgent action is needed to reverse this trend and safeguard the remaining biological heritage. Restoration and conservation efforts are pivotal in this endeavour, aiming to rehabilitate degraded ecosystems, protect vulnerable species, and promote sustainable land and resource management practices (CBD 2018). Effective conservation strategies require a multi-faceted approach, integrating scientific research, policy interventions, community engagement, and public awareness campaigns. Moreover, concerted international cooperation is essential to address transboundary conservation challenges and ensure the equitable distribution of conservation benefits (UNEP 2020).

Understanding Biodiversity

To formulate effective strategies for restoring and safeguarding biodiversity, it is imperative to understand its various dimensions, the ecosystem services it provides, and the underlying factors contributing to its loss.

Components of Biodiversity

Biodiversity encompasses three main components: species diversity, genetic diversity, and ecosystem diversity (CBD 2014). Species diversity pertains to various organisms within a given ecosystem, including plants, animals, fungi, and microorganisms. Genetic diversity refers to the variability of genes among populations of the same species, enabling adaptation to changing environmental conditions. Ecosystem diversity encompasses the range of habitats, communities, and ecological processes across landscapes, ranging from forests and wetlands to grasslands and coral reefs (Gogoi et al. 2024a, b).

Ecosystem Services Provided by Biodiversity

Biodiversity underpins a broad spectrum of ecosystem services essential for human well-being (MEA 2005). These services encompass provisioning services like food, fresh water, and medicinal resources; regulating services such as climate regulation, pest control, and water purification; cultural services such as spiritual and recreational benefits; and supporting services like soil formation, nutrient cycling, and pollination. The loss of biodiversity jeopardizes the resilience of ecosystems and compromises their ability to provide these essential services (MEA 2005).

Factors Contributing to Biodiversity Loss

Several interconnected factors contribute to the ongoing loss of biodiversity. Anthropogenic activities have dramatically accelerated biodiversity decline, surpassing natural extinction rates by orders of magnitude (Fig. 16.1). Habitat destruction and fragmentation, driven by activities such as deforestation, urbanization, and agricultural expansion, are among the primary drivers of species extinctions (Wilcove et al. 2013). Habitat destruction ranks among the leading drivers of biodiversity loss worldwide. Activities such as deforestation, urbanization, agriculture, and infrastructure development lead to the degradation and fragmentation of natural habitats, diminishing species' survival chances (Fig. 16.1). For instance, the conversion of forests for agricultural purposes has led to the loss of critical habitats for numerous species, including endemic and endangered ones (Laurance et al. 2014). Pollution, including air and water pollution, disrupts ecosystems and poses health risks to wildlife and humans (Sala et al. 2000). For instance, chemical runoff from agricultural activities has been implicated in the decline of amphibian population's worldwide (Hayes et al. 2010). Figure 16.1 shows key contributors to biodiversity.

Climate change exacerbates biodiversity loss by modifying habitats, disrupting species distributions, and amplifying the frequency and intensity of extreme weather events (IPCC 2019). Including altered temperature regimes, precipitation patterns, and sea level rise these changes disrupt ecosystems and species' distributions, leading to habitat loss, shifts in species' phenology, and increased vulnerability to pests and diseases (IPBES 2019a, b). Coral bleaching events, attributed to rising sea temperatures, have devastated marine biodiversity in coral reef ecosystems (Hughes et al. 2017). Overexploitation of natural resources, such as overfishing and poaching, threatens the survival of numerous species (Hogue and Breon, 2022). Unsustainable harvesting practices deplete populations beyond their capacity for recovery, disrupting ecological dynamics and food webs. The overhunting of large mammal

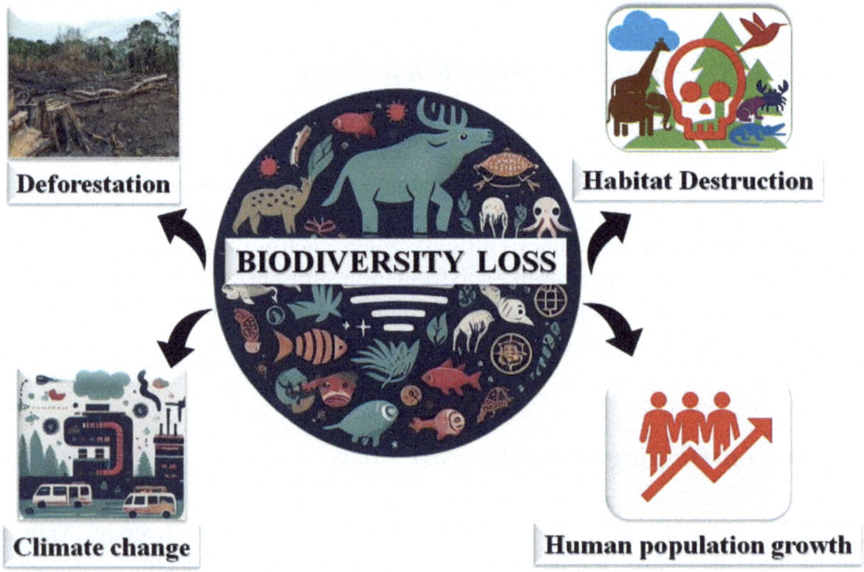

Fig. 16.1 Key contributors to biodiversity loss in an ecosystem

species, such as tigers and rhinoceroses, has pushed them to extinction (Ripple et al. 2016).

Moreover, introducing invasive species substantially threatens native biodiversity by surpassing or predating native species (Vitousek et al. 1997). It can harm native biodiversity by outcompeting native species, predation, habitat alteration, and transmission of diseases. The brown tree snake's introduction to Guam led to the extirpation of numerous bird species and altered forest dynamics (Savidge 1987).

Restoring Biodiversity

Ecological Restoration Principles

Restoration is gaining global acceptance as an effective approach to addressing various environmental issues. The United Nations (UN) General Assembly has officially designated the 2021–2030 as the UN Decade on Ecosystem Restoration. In December 2022, global leaders endorsed the Kunming-Montreal Global Biodiversity Framework. The Framework outlines goals for the year 2050 that will be achieved by implementing defined and quantifiable targets by 2030 to preserve and restore the natural ecosystem (Shackelford and McDougall 2023).

Ecological restoration is motivated by a core set of principles to revitalize and enhance degraded ecosystems to their initial state. These notions serve as a framework for designing and implementing restoration projects (Shackelford and McDougall 2023). The major objective of restoration efforts should be to prioritize ecosystem integrity and safeguard and enhance the ecological well-being of the target ecosystem.

When formulating plans for restoration projects, it is essential to take a holistic approach and consider various factors, including biological succession, the effects of climate change, and the requirements for ongoing maintenance. Restoration efforts should acknowledge and improve ecological services, including clean water, carbon storage, and recreation. Ecological services may assist in prioritizing restoration initiatives that maximize social benefits. By adhering to these principles, ecological restoration programs can effectively restore impaired ecosystems, promote biodiversity conservation, and enhance the resilience of natural landscapes to environmental stresses.

Case Studies of Successful Ecological Restoration

Successful restoration projects can act as examples, showing how different methods and strategies can successfully restore ecological balance, biodiversity, and ecosystem services. Following is some of the successful ecological restorations in Northeast India.

Deepor Beel

Deepor Beel covers an area of almost 40 km^2 and functions as a natural reservoir for runoff in Guwahati. The wetland sustains a diverse range of plant and animal life, In addition to its biological importance, Deepor Beel supports the livelihoods of local populations by providing valuable resources such as fish, fodder, and water for agriculture. The wetland restores underground water sources and lessens the effects of floods, underscoring the need to protect it for the health of the ecosystem and people's well-being (Kumar et al, 2020). Deepor Beel is subject to numerous human-induced stressors. The marsh has been encroached upon for residential and commercial development due to the fast urbanization of surroundings and the expansion of Guwahati. The wetland's area has been drastically diminished, and its hydrology has been altered due to the presence of illegal settlements and unplanned infrastructure construction. The disposal of municipal solid trash near the wetland intensifies the levels of pollution. The excessive exploitation of resources such as fish and plants has caused degradation to the ecosystem—The proliferation of invasive plant species such as water hyacinths has also obstructed water bodies, hence impacting indigenous flora and wildlife (Hoque and Sharma 2020).

In response to the need to restore Deepor Beel, government institutions, non-governmental organizations (NGOs), and local communities have implemented several efforts. These initiatives include a variety of actions that are focused on restoring the ecological equilibrium and guaranteeing the sustainable utilization of the wetland's resources. The execution of pollution control measures in Guwahati includes the development of sewage treatment plants (STPs) to treat wastewater before it enters the beel (wetland), as well as the enforcement of trash segregation and recycling programs in neighbouring areas. Habitat restoration initiatives encompass the act of reforesting the wetland by strategically planting indigenous tree species to establish buffer zones. Additionally, the eradication of invasive species such as water hyacinths is undertaken to reinstate open water regions. Community engagement is an essential element, involving the implementation of awareness initiatives to enhance understanding of the significance of Deepor Beel. conservation initiatives have resulted in a rise in the number of diverse avian and aquatic species (Hoque and Sharma 2020).

Restoration of Hargila Storks

The ecological restoration of the hargila bird, commonly known as the Greater Adjutant Stork, in Assam, serves as an impressive illustration of conservation efforts led by the local population. The hargila, a species at risk of extinction, has had a substantial recovery as a result of the committed endeavours spearheaded by conservation scientist Dr. Purnima Devi Barman and the "Hargila army," a predominantly female group consisting of local residents. The population of the Greater Adjutant stock has significantly declined because of habitat degradation caused by urbanization and negative opinions held by residents. To address this issue, Dr. Barman concentrated on altering these perspectives by incorporating the bird into the local culture and customs. The initiatives encompassed the coordination of community events like baby showers for the storks, integration of hargila motifs into traditional Assamese textiles, and dissemination of knowledge within the community regarding the ecological significance of the birds. These endeavours have achieved remarkable success. The number of nests in the Kamrup district has witnessed a significant rise, surging from 27 in 2007 to more than 210. Also, the Hargila Army, consisting of more than 400 women, plays an essential part in overseeing and safeguarding nesting places, healing injured birds, and raising awareness. The community-driven approach offers a robust framework for conservation that integrates ecological and social advantages (Source: onlifeandwildlife.com).

Conserving Biodiversity

The dichotomy between human development and environmental protection arises from interconnected issues, including rapid population growth, overexploitation, and habitat degradation (Plieninger et al. 2014). Preserving species variety is critical to ecological balance and resilience. Each species has a distinct role in its ecosystem, contributing to the nutrient cycle, pollination, and pest management.

Protected Areas and Reserves

Protected Areas and Reserves are designated zones where human activity is regulated, managed, or restricted to varying degrees to preserve natural habitats, species, and ecological processes (Ghosh-Harihar et al. 2019). These areas encompass diverse landscapes, ranging from terrestrial ecosystems like forests, grasslands, and deserts to marine environments such as coral reefs, mangroves, and deep-sea habitats (Zhang et al. 2022).

The primary goal of Protected Areas and Reserves is biodiversity conservation, which entails the protection of species, habitats, and genetic diversity (Mackey et al. 2008). By safeguarding critical habitats and ecosystems, these areas provide refuge for endangered species, support population recovery, and maintain ecological balance. Clean water, air purification, climate regulation, and nutrient cycling are just a few of the important ecosystem services they provide, and they also help ecosystems be more resilient in the face of environmental hazards. However, their effectiveness hinges on concerted efforts to address conservation challenges, uphold legal protections, and foster inclusive governance frameworks prioritizing ecological integrity, social equity, and cultural diversity.

Biodiversity Hotspots

Biodiversity hotspots represent regions of extraordinary ecological richness and high species endemism that face significant threats from human activities (Mittermeier et al. 2011). These areas, identified and delineated by conservation scientists, serve as focal points for biodiversity conservation efforts due to their immense biological diversity and the urgency of their conservation needs (Myers et al. 2000).

Biodiversity hotspots are characterised by a combination of factors, including high levels of species diversity, high rates of endemism (species found nowhere else), and severe habitat loss and degradation (Marchese 2015). Despite occupying less than 2% of Earth's surface, these hotspots hold a disproportionately large number of plant and animal species, resulting in vast biodiversity reservoirs (Mittermeier et al. 2011). Biodiversity and ecological services like pollination, water purification, climate regulation, and disease management depend on protecting these hotspots.

Community-Based Conservation Initiatives

Community-based conservation initiatives represent a paradigm shift in biodiversity conservation, emphasizing the integral role of local communities in stewarding natural resources and protecting biodiversity (Armitage et al. 2020). The purpose of these initiatives is to promote sustainable management techniques that are beneficial to both people and the environment.

(a) Participatory Planning and Decision-making: From identifying conservation priorities and setting objectives to designing and implementing management strategies. The participatory method guarantees that conservation actions are locally suitable, socially acceptable, and meet local stakeholders' needs and goals.
(b) Capacity Building and Empowerment: This may involve training in sustainable resource management practices, supporting the establishment of community-based organizations, and fostering leadership skills among community members.
(c) Inclusive Governance and Co-management: Co-management arrangements, where responsibilities and decision-making authority are shared between communities and external partners, foster collaborative management approaches that leverage complementary strengths and perspectives.
(d) Alternative Livelihoods and Economic Incentives: Recognizing the linkages between poverty and environmental degradation, community-based conservation initiatives seek to address socio-economic drivers of biodiversity loss by providing alternative livelihood opportunities that reduce dependence on unsustainable natural resource extraction. This may involve supporting eco-tourism initiatives, sustainable agriculture practices, non-timber forest product

harvesting, or small-scale enterprises that generate income while promoting environmental conservation.

(e) Cultural Revitalization and Traditional Knowledge: Community-based conservation initiatives celebrate and preserve local cultures, traditions, and indigenous knowledge systems that are often deeply intertwined with ecosystems and biodiversity.

Policy and Practice Implications

Strategies for Effective Implementation at Local, National, and Global Levels

Effective implementation of biodiversity conservation strategies requires a multifaceted approach tailored to various levels of governance. Engaging local communities in the decision-making process ensures that conservation strategies are contextually relevant and socially acceptable. This approach empowers local populations to manage their natural resources sustainably and can lead to successful conservation outcomes.

National strategies should involve inclusive governance frameworks that include local communities, indigenous peoples, government agencies, and NGOs. Establishing and enforcing legal protections for critical habitats and species is essential. National policies should also support sustainable funding for conservation projects and integrate biodiversity considerations into broader land-use planning and development policies. Global biodiversity conservation requires international cooperation, exemplified by agreements such as the Kunming-Montreal Global Biodiversity Framework. These frameworks outline global goals and targets, promoting coordinated efforts to preserve and restore natural ecosystems. Global strategies must consider the impacts of climate change on biodiversity and integrate adaptive measures to enhance the resilience of ecosystems.

Role of Governments, NGOs, and Communities in Conservation Efforts

Governments play a crucial role in formulating and enforcing policies that protect biodiversity. This includes establishing protected areas, regulating resource extraction, and implementing pollution control measures. Governments can provide financial support for conservation projects and research, ensuring sustainable funding for long-term conservation efforts. NGOs raise awareness about biodiversity issues and advocate for stronger conservation policies. Communities can integrate traditional knowledge and cultural practices into conservation efforts, enhancing both ecological and cultural resilience.

Economic and Social Benefits of Biodiversity Restoration

Biodiversity restoration enhances ecosystem services such as clean water, carbon storage, and recreation. These services are vital for human well-being and can provide significant economic benefits by reducing costs associated with water purification, flood control, and climate regulation. Restoration projects can create new livelihood opportunities for local communities, such as eco-tourism, sustainable agriculture, and the harvesting of non-timber forest products (Singh et al. 2022). These economic incentives can reduce dependence on unsustainable resource extraction and promote long-term conservation goals. Biodiversity restoration can also strengthen cultural identity and social cohesion (Kumar and Choudhury 2021).

Recommendations

For ecological restoration to be successful, a comprehensive approach is required, which begins with careful planning and assessment. It is necessary to have a thorough grasp of the environment's historical and current state, as well as to establish attainable goals. It is essential to prepare the site, with a particular emphasis on eradicating invasive species and maintaining the health of the soil. The utilization of a wide variety of native species and the restoration of natural hydrology should be the implementation's top priorities. Continuous monitoring and adaptive management are necessary. When tracking progress and adjusting strategies are as required increasing support and stewardship can be accomplished by involving local communities through education, volunteer initiatives, and stakeholder involvement. In addition, incorporating scientific research and new practices is a way to improve restoration processes. For success over the long run, it is essential to ensure that supportive policies and sustainable funding are in place. Further strengthening restoration efforts can be strengthened by considering adaptation to climate change, biodiversity conservation, and cultural values. This will ultimately result in ecosystems that are resilient and thriving.

Conclusion

This study underscores the pivotal role of community-based conservation initiatives. By involving local communities in decision-making processes, these initiatives leverage traditional knowledge and cultural practices, fostering a sense of ownership and responsibility towards natural resources. This participatory approach ensures that conservation strategies are contextually relevant and socially acceptable, aligning with the needs and aspirations of local stakeholders. Furthermore, the study emphasizes capacity building and empowerment, enabling communities to manage natural resources sustainably and implement practical conservation activities. Adopting innovative techniques such as genetic rescue and assisted migration offers promising avenues for bolstering populations and enhancing genetic diversity in endangered species. It also highlights the success of community-driven initiatives, such as marine protected areas in the Philippines and community forestry programs in Nepal, demonstrating local stewardship's effectiveness in restoring and protecting biodiversity. These case studies provide a blueprint for scalable and replicable models of conservation. By fostering collaboration across governments, NGOs, academia, industry, and civil society, and by nurturing a reverence for the natural world, we can aspire towards a future where biodiversity thrives, ensuring the sustainability of our planet for generations to come.

References

Armitage D, Mbatha P, Muhl EK, Rice W, Sowman M (2020) Governance principles for community-centered conservation in the post-2020 global biodiversity framework. Conserv Sci Pract 2(2):e160

CBD (2014) Global biodiversity outlook 4. Secretariat of the Convention on Biological Diversity

CBD (2018) Global biodiversity outlook 5. Secretariat of the Convention on Biological Diversity

Ghosh-Harihar M, An R, Athreya R, Borthakur U, Chanchani P, Chetry D, Datta A, Harihar A, Karanth KK, Mariyam D (2019) Protected areas and biodiversity conservation in India. Biol Cons 237:114–124

Gogoi N, Sarma A, Choudhury M, Samanta P, Faizan M, Sadhak S (2024a) Soil heavy metal pollution and ecological risk assessment in disturbed and undisturbed soil of Morigaon, Assam. Discov Environ 2(1):13. https://doi.org/10.1007/s44274-024-00039-z

Gogoi N, Choudhury M, Ul Hasan MS, Changmai, B, Baruah D, Samanta P (2024) Soil organic carbon pool in diverse land utilization patterns in North-East India: an implication for carbon sequestration. Environ Dev Sustain, 1–27. https://doi.org/10.1007/s10668-024-05037-y

Hayes TB et al (2010) Atrazine induces complete feminization and chemical castration in male African clawed frogs (*Xenopus laevis*). Proc Natl Acad Sci 107(10):4612–4617

Hoque NA, Sharma A (2020) Sustainable management and farmer's willingness to participate: a case study of Deepor Beel. Ecol Environ Conserv 26(4):1806–1815

Hogue A S, Breon K (2022) The greatest threats to species, Conservation Science and Practice, 4(5), e12670

Hughes TP et al (2017) Global warming and recurrent mass bleaching of corals. Nature 543(7645):373–377

IPBES (2019). Summary for policymakers of the global assessment report on biodiversity and ecosystem services of the Intergovernmental Science-Policy Platform on Biodiversity and Ecosystem Services

IPBES (2019) Global assessment report on biodiversity and ecosystem services. Intergovernmental Science-Policy Platform on Biodiversity and Ecosystem Services

IPCC (2019) IPCC special report on climate change and land. Intergovernmental Panel on Climate Change

Kumar D, Choudhury M, Rathore A (2020) Valuation of ecosystem services and benefits of Son Beel wetland in Assam, India: A case study of natural solutions to climate change & water: the worldwide online climate change conference. Hamburg University, Germany

Kumar D, Choudhury M (2021). Recognizing economic values of wetland ecosystem services: a study of emerging role of monetary evaluation of Chandubi ecosystem and biodiversity. In:

Wetlands conservation: current challenges and future strategies, pp 87–110. https://doi.org/10.1002/9781119692621.ch5
Laurance WF et al (2014) Averting biodiversity collapse in tropical forest protected areas. Nature 489(7415):290–294
Mackey BG, Watson JE, Hope G, Gilmore S (2008) Climate change, biodiversity conservation, and the role of protected areas: an Australian perspective. Biodiversity 9(3–4):11–18
Maestro M, Pérez-Cayeiro ML, Chica-Ruiz JA, Reyes H (2019) Marine protected areas in the 21st century: current situation and trends. Ocean Coast Manag 171:28–36
Marchese C (2015) Biodiversity hotspots: a shortcut for a more complicated concept. Glob Ecol Conser 3:297–309
MEA (2005) Millennium ecosystem assessment: ecosystems and human well-being: synthesis. Island Press
Mittermeier RA, Turner WR, Larsen FW, Brooks TM, Gascon C (2011) Global biodiversity conservation: the critical role of hotspots. In Biodiversity hotspots: distribution and protection of conservation priority areas. Springer, pp 3–22
Myers N, Mittermeier RA, Mittermeier CG, Da Fonseca GA, Kent J (2000) Biodiversity hotspots for conservation priorities. Nature 403(6772):853–858
Plieninger T, Van der Horst D, Schleyer C, Bieling C (2014). Sustaining ecosystem services in cultural landscapes. Ecol Soc 19(2)
Ripple WJ et al (2016) Saving the world's terrestrial megafauna. Bioscience 66(10):807–812
Sala OE et al (2000) Global biodiversity scenarios for the year 2100. Science 287(5459):1770–1774
Savidge JA (1987) Extinction of an island forest avifauna by an introduced snake. Ecology 68(3):660–668
Shackelford N, McDougall C (2023) Ecosystem restoration, regeneration and rewilding. BMC Ecol Evo 23:52
Singh S, Singh P, Sharma A, Choudhury M (eds) (2022) Agriculture waste management and bioresource: the circular economy perspective. Wiley. https://onlinelibrary.wiley.com/doi/book/10.1002/9781119808428
UN Environment Programme (UNEP) (2020) Global biodiversity outlook 5. United Nations Environment Programme
Vitousek PM et al (1997) Human domination of Earth's ecosystems. Science 277(5325):494–499
Wang S, Adhikari K, Wang Q, Jin X, Li H (2018) Role of environmental variables in the spatial distribution of soil carbon (C), nitrogen (N), and C:N ratio from the northeastern coastal agroecosystems in China. Ecol Ind 84:263–272
Wilcove DS et al (2013) Quantifying threats to imperiled species in the United States. Bioscience 48(8):607–615
Zhang Z, Ge S, Fan L-C, Guo S, Hu Q, Ahammed GJ, Yan P, Zhang L-P, Li Z-Z, Zhang J-Y (2022) Diversity in rhizospheric microbial communities in tea varieties at different locations and tapping potential beneficial microorganisms. Front Microbiol 13:1027444